AUTOMATION IN AGRICULTURE - SECURING FOOD SUPPLIES FOR FUTURE GENERATIONS

Edited by **Stephan Hussmann**

Automation in Agriculture - Securing Food Supplies for Future Generations
http://dx.doi.org/10.5772/intechopen.69016
Edited by Stephan Hussmann

Contributors

Redmond Ramin Shamshiri, Ibrahim A. Hameed, Manoj Karkee, Cornelia Weltzien, Jehoon Sung, Abdullah Ansari, Vasnie Ganaeshnauth, Sirpaul Jaikishun, Oudho Homenauth, Florian Johannes Knoll, Vitali Czymmek, Jesús Raúl Martinez, Ernesto Martinez, Miguel Martinez, Manuel Miranda, Humberto Cervantes, Takashi Okayasu, Takehiko Hoshi, Ken-Ichiro Yasuba, Hideto Kurosaki, Ivan Mašín, Zhihong Zhang, Xiaoyang Wang, Qinghui Lai, Zhaoguo Zhang, Ibrahim Mohammed Gana, Gbabo Agidi

Published by InTech
Janeza Trdine 9, 51000 Rijeka, Croatia

Publishing Process Manager Martina Usljebrka
Technical Editor SPi Global
Cover InTech Design team

First published March, 2018
Printed in Croatia
Legal deposit, Croatia: National and University Library in Zagreb

Additional hard copies can be obtained from orders@intechopen.com

Automation in Agriculture - Securing Food Supplies for Future Generations, Edited by Stephan Hussmann
p. cm.
Print ISBN 978-953-51-3873-0
Online ISBN 978-953-51-3874-7

For EU product safety concerns:
IN TECH d.o.o., Prolaz Marije Krucifikse Kozulić 3, 51000 Rijeka, Croatia,
info@intechopen.com or visit our website at intechopen.com.

Meet the editor

Stephan Hussmann received his ME and PhD degrees from the University of Siegen, Siegen, Germany, in 1995 and 2000, respectively. From 1996 to 2000, he was a research associate in the Center for Sensor Systems (ZESS), University of Siegen, and a development engineer in Aicoss GmbH, Siegen. From 2001 to 2003, he was a lecturer in the Department of Electrical and Computer Engineering, University of Auckland, Auckland, New Zealand. Since the end of 2004, he has been a professor in the Faculty of Engineering, West Coast University of Applied Sciences (FHW), Heide, Germany in the area of microprocessor technology and electronic systems. He has been widely consulted in the industry and has more than 80 publications, which include book chapters, international patents, and refereed journals and conference proceedings or papers. His research interests include low-cost multisensor system design, real-time 2D/3D image processing, embedded system design, machine vision, and agricultural automation.

Contents

Preface

According to Prof. Dickson Despommier, by the year 2050, nearly 80% of the earth's population will reside in urban centers. Applying the most conservative estimates to current demographic trends, the human population will increase by about 3 billion people during the interim. An estimated 109 hectares of new land (about 20% more land than is represented by the country of Brazil) will be needed to grow enough food to feed them, if traditional farming practices continue as they are practiced today. At present, throughout the world, over 80% of the land that is suitable for raising crops is in use (sources: FAO and NASA). Historically, some 15% of that has been laid waste by poor management practices. What can be done to avoid this impending disaster?

Emeritus Prof. Dickson Despommier has conducted laboratory-based biomedical research at Columbia University in the USA for 27 years. He said that one possible solution is indoor farming. However, not all crops can easily be moved in an indoor environment. Nevertheless, to secure the food supply for future generations, it is necessary to increase the automation level in agriculture significantly either for indoor or for outdoor applications. This book intends to provide the reader with a comprehensive overview of the current state-of-the-art automation in agriculture, featuring an easy-to-follow, vignette-based format that focuses on the most important evidence-based developments in this critically important area.

The book comprises two main sections. In the first section, the impact of the Fourth Industrial Revolution and Trends of Engineering System Evolution on agriculture is described. The Fourth Industrial Revolution will send a ripple effect of far-reaching repercussions throughout the labor-intensive field of agriculture. Dr. J. Sung illustrates this effect in Chapter 1. In Chapter 2, my PhD students, F. Knoll (MSc) and V. Czymmek (MSc), explain the influence of the Fourth Industrial Revolution on the field of organic farming. In our research group at the West Coast University of Applied Science, for example, a robot is being researched that automatically removes the weeds in an organic farm. In Chapter 3, Associate Prof. I. Mašín focuses on the so-called trends of engineering system evolution. He describes natural transitions of the engineering system from one state to another, and they are generally valid for all engineering disciplines. However, he relates this method to agricultural technology. The second section contains industrial automation examples in agriculture. Chapter 4, written by Dr. Z. Zhang, gives a review on variable-rate sprayer applications based on real-time sensor technologies. Variable-rate spraying of the canopy, for example, allows growers to apply an adjusted volume rate of pesticides to the target, based on canopy size, and to apply plant protection products in an economical and environmentally sound manner. In the next chapter (Chapter 5), Dr. R. R. Shamshiri reports on designing a simulation and control platform for experimenting with sensors and manipulators in robotic harvesting of sweet pepper. The objective was to provide a completely simulated environment for improvement of visual serving tasks through easy testing and debugging of control algorithms with zero damage risk to the real robot and to the actual equipment. Associate Prof. T. Okayasu discusses in his chapter (Chapter 6) a self-fabricated ubiquitous environment control system (UECS). The flexibility and concept of the developed UECS have been very effective to enable sophisticated environmental control technology to be applied to small- and medium-scale greenhouses in Japan. In Chapter 7, Dr. J. Martinez describes the use of machine vision systems for automatic color analysis in agriculture. It could be said that machine vision systems are appropriate to improve the actual agricultural systems making them more useful, efficient, practical, and reliable. The research work of Dr. I. Gana, illustrated in Chapter 8, focuses on the design and the construction of an automatic system for grain beverage processing. The automated system allows an efficient workflow, reduces human labor, and ensures safety and a hygiene production process. In the last chapter (Chapter 9), Prof. A. Ansari points out the effect of vermicompost and other fertilizers on the growth and productivity of pepper plants in Guyana.

I would like to extend my most sincere gratitude to the authors who have generously contributed chapters to this book, without whom this project would not have been a success. Furthermore, I would like to thank the West Coast University of Applied Sciences in Heide, Germany, for approving my sabbatical. Otherwise, I would not have been able to edit this book. Also, I would like to give my heartfelt thanks to InTech Publishers, and I look forward to working hopefully with them in many more projects in the future as well. Last but not the least, my appreciation goes to Ms. Martina Usljebrka, the Publishing Process Manager assigned to this book, who has rendered her utmost support in putting the materials together.

Prof. Dr.-Ing. Stephan Hussmann
Faculty of Engineering
West Coast University of Applied Sciences
Heide, Germany

Section 1

The 4th Industrial Revolution and Trends of Engineering System Evolution in Agriculture

Chapter 1

The Fourth Industrial Revolution and Precision Agriculture

Jehoon Sung

Additional information is available at the end of the chapter

http://dx.doi.org/10.5772/intechopen.71582

Abstract

The Fourth Industrial Revolution will see the convergence of artificial intelligence and data technology as a new solution to address industrial and social problems across the globe, by integrating cyber and physical fields. The Fourth Industrial Revolution will send a ripple effect of far-reaching repercussions throughout the labor-intensive field of agriculture. Combining artificial intelligence and big data will evolve into a high-tech industry that operates itself. These technologies allow for precision agriculture, such as yield monitoring, diagnosing insect pests, measuring soil moisture, diagnosing harvest time, and monitoring crop health status. In particular, the Internet of things (IoT) will measure the temperature, humidity, and amount of sunlight in production farms, making it possible for remote control via mobile devices. It will not only boost the production of the farms but also add to their value.

Keywords: Fourth Industrial Revolution, precision agriculture, sensing

1. Introduction

The key phrase used at the World Economic Forum (WEF)[1] was the Fourth Industrial Revolution. Klaus Schwab, founder of the WEF, argued that the Fourth Industrial Revolution has already arrived. He argued that the pace, scope, and influence of social changes that follow the Fourth Industrial Revolution will be entirely different from previous revolutions.

[1]The World Economic Forum, better known as the Davos Forum, is a Swiss nonprofit foundation, based in Cologny, Geneva. Recognized by the Swiss authorities as an international body, its mission is cited as "committed to improving the state of the world by engaging business, political, academic, and other leaders of society to shape global, regional, and industry agendas." The Forum is best known for its annual meeting at the end of January in Davos, a mountain resort in Graubünden, in the eastern Alps region of Switzerland (from Wikipedia, accessed July 30, 2017) [1].

The Fourth Industrial Revolution, or 4IR, refers to the oncoming revolutionary era in which information and communication technology (ICT) will converge. The revolution will spark new technological innovations in six areas: artificial intelligence, robotics, Internet of things (IoT), unmanned vehicles, three-dimensional printing, and nanotechnology. The 4IR will include a variety of new technologies that use big data to incorporate the physical, biological, and digital worlds in a way that will affect all sectors of life.

One example of the 4IR is online to offline, or O2O, which integrates the physical and digital worlds. O2O can use smart watches that obtain real-time information from patients and confer it to integrated computer data. Other examples of the 4IR include virtual reality (VR) and augmented reality (AR).

4IR technologies have the potential to connect billions on the web, dramatically improve business organizational efficiency, and improve the natural environment through improved asset management.

The 4IR will become a new innovative division of life that will replace human intelligence and wisdom, combining artificial intelligence with robotic technology as a substitute for labor.

General Electric (GE) is a typical example of the 4IR occurring in the present. GE, originally a lighting company, has merged into the domains of electrical equipment, televisions, computers, home appliances, generators, and even medical equipment and aircraft engines. Already successful in previous fields, GE is now the top aerospace manufacturer as well.

Adopting the concept of the 4IR, GE created a new revenue model that surpassed the sales of all other aircraft engine manufacturers. Mounting sensors on aircraft engines was their key to success. The in-flight sensors connect to ground data centers and send more than 300 different values of real-time information to and from the aircraft. Data sent includes engine conditions, weather conditions, and fuel efficiency. These transmissions allow the ground centers to analyze the data and return an optimized flight path to the aircraft in real time, reducing fuel usage and saving an estimated two billion dollars per year. In addition, the sensors monitor the safety status of the aircraft in real time, anticipating abnormal conditions and dramatically reducing accidents and inspection costs, allowing airlines to improve security as well as operational safety.

The 4IR is developing in every sector of life, not only in telecommunications, automobiles, energy, manufacturing services, security, and bioenergy but also in the fields of medicine and robotics. The 4IR is now being commercialized in a variety of endeavors, including the Google Car, Amazon's Drone Delivery system, and Dr. Watson: an AI doctor. One of the ways in which the 4IR is expected to approach new problems is in the field of agriculture.

National policies related to the 4IR, based on global trends, are being implemented across the planet. The following chapter foretells changes coming to agriculture and preparations required in the field of science and technology in regard to the 4IR.

2. The Fourth Industrial Revolution and agriculture

Over 200 years ago, more than 90% of Earth's population was engaged in agriculture, but now more than 80% of the populations of OECD major countries are engaged in the service industry. The population engaged in agriculture, at present, is merely 2–3%. Not only has the population involved in agriculture been reduced, in most developed countries, the age of individuals in farming households is increasing as well. In the Republic of Korea, more than 50% of the population of farm households is over 60 years old, and over 40% is over 65. The population of workers around the globe has shifted from agriculture to manufacturing and manufacturing service industries. Thus, in the current world economy, only 5 % of the world's population works in agriculture, yet it accounts for more than 60% of the world's business [2].

Accepting this reality, developed countries such as the USA and Japan are trying to solve agricultural issues through mechanization, automation, and modernization. The 4IR will serve as the opportune time to accelerate the scale and commercialization of agriculture.

In response to this trend, future agriculture is expected to evolve into high-tech industries where systems are coupled with artificial intelligence[2] and big data[3]. The systems will converge into a single unit in which farm machinery, seeding the soil, farm management, production forecasting, and irrigation are combined. Using the core technology of the 4IR, robots, big data, and AI will combine with agriculture to create a new era of superfusion. The era will evolve multifaceted economic, social, and ethical values fused with various industries and expressed in business models [3].

There are three means by which the 4IR will have a major impact on the agricultural sector. First, precise optimization will solve many current problems in agriculture. Agriculture is a representative industry in which inputs and outputs are inconsistent. In terms of worldwide food production, enough food is produced for the entire population, yet 30–50% of produced food is discarded, while many die of starvation. About 80% of the water on the planet is used for agriculture, yet only 20% of viable crop is grown, and the remaining unused surplus is discarded. In the UK, the use of nitrogen fertilizer resulted in blue disease. Each of these problems can be solved via precision agriculture. Precision agriculture,[4] a method by which

[2]Artificial intelligence (AI) is intelligence exhibited by machines. In computer science, the field of AI research defines itself as the study of "intelligent agents": any device that perceives its environment and takes actions that maximize its chance of success at some goal. Colloquially, the term "artificial intelligence" is applied when a machine mimics "cognitive" functions that humans associate with other human minds, such as "learning" and "problem solving." As machines become increasingly capable, mental facilities once thought to require intelligence are removed from the definition (from Wikipedia, accessed 30-07-2017).

[3]Big data is a term for data sets that are so large or complex that traditional data processing application software is inadequate to deal with them. Big data challenges include capturing data, data storage, data analysis, search, sharing, transfer, visualization, querying, updating, and information privacy. Lately, the term "big data" tends to refer to the use of predictive analytics, user behavior analytics, or certain other advanced data analytics methods that extract value from data and seldom to a particular size of data set (from Wikipedia, accessed 30-07-2017).

[4]Precision agriculture (PA) or satellite farming or site-specific crop management (SSCM) or precision agriculture is a farming management concept based on observing, measuring, and responding to inter and intra-field variability in crops. The goal of precision agriculture research is to define a decision support system (DSS) for whole farm management with the goal of optimizing returns on inputs while preserving resources (from Wikipedia, accessed July 30, 2017).

growth and soil conditions are calculated in order to accurately manage crops, can solve the problem by constructing an optimized agricultural system that connects production, distribution, and consumption.

Second, the reversion of rural production elements, including human resources, will have a major impact on agriculture. Capital, labor, and technological resources that left farming villages in previous generations are likely to return during the 4IR. This is because the workforce of cities will find that rural areas provide the only labor that gives time for rest and relaxation.

Third, 4IR technologies will have a significant impact on weather-related problems. Agriculture is heavily affected by the weather, and currently science has no means by which to accurately predict and control it. Hence, we say that we are fellow farmers with God. For this reason, farming is highly dependent on intelligence and wisdom, including human experience, and thus it is difficult to standardize. 4IR technology can make decisions that surpass human wisdom and experience. It will solve certain problems that cannot be solved with current technology, such as livestock odors, the cost of too much processing, and the likelihood of pest occurrence due to climate change. So, the 4IR can be seen as an "agro-friendly" revolution, unlike our current revolution. At the same time, it will lead to greater technological innovations and far-reaching changes throughout the economy, society, and life.

3. The Fourth Industrial Revolution and changes in agriculture

The agriculturally friendly Fourth Industrial Revolution will expand the scope of agriculture in various fields, such as culture, welfare, and healing in production-oriented agriculture. As shown in **Figure 1**, the 4IR will lead to a greater amount of communal and independent cultivation through cultural activities, such as combining agriculture with games and leisure, human welfare agriculture in the age of aging, and agricultural activities with plants and animals [4].

The expansion of agriculture through the 4IR is expected to vary greatly in the fields of production, distribution, and consumption.

3.1. Production of agricultural products

Changes in agricultural production in the 4IR will occur primarily in agricultural facilities with smart farming technology. In capable facilities, controlling the growth environment will add to the value of agricultural products. In Korea, three stages must be completed in order to promote smart farms in agricultural facilities. The first stage, completed prior to 2017, is the convenience improvement stage. In this stage, facilities were upgraded to allow farmers to check the growth status of agriculture via mobile devices. Thus, farmers do not need to travel to farms for menial tasks such as temperature control. The second stage, which is expected to be completed by 2020, is productivity improvement. In this stage, profits are increased through precise control and optimal prescription of agriculture. The

http://dx.doi.org/10.5772/intechopen.71582

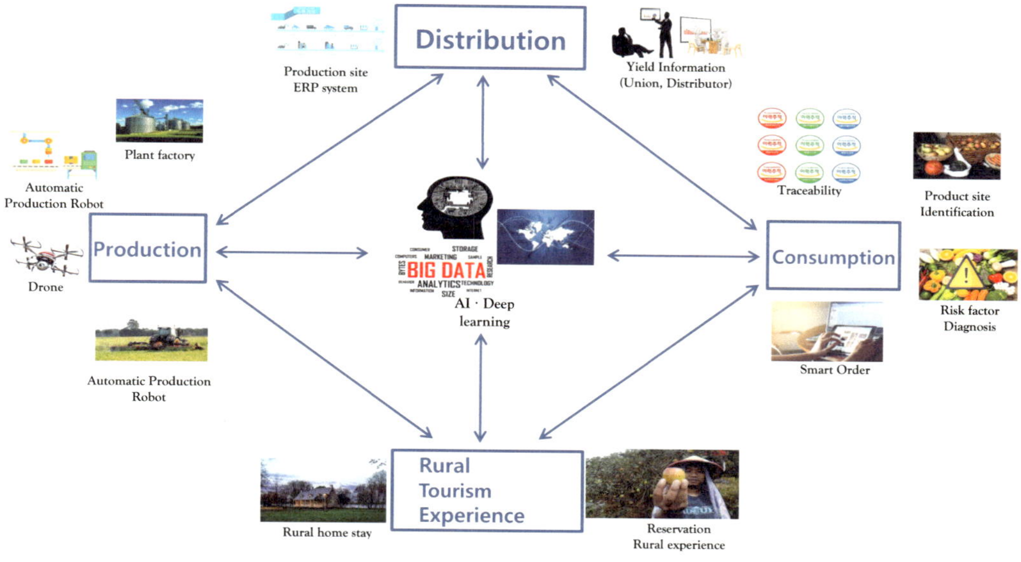

Figure 1. Illustration of the future agriculture with the Fourth Industrial Revolution (source: Fourth Industrial Revolution and Agriculture, Korea institute of planning and evaluation for technology in food, agriculture and forestry, 2016) [4].

third stage is the completion stage, in which all of the facility conditions are automated according to the growth conditions of the crop based upon the crop's growth model. The Korea Rural Development Administration provides a platform[5] for testing various sensors and technologies in smart farms, in order to help farmers quickly and efficiently move through the three stages.

As shown in **Figure 2**, the 4IR will also make a big difference in open-field agriculture. There are three stages in which this technology can be used: monitoring the area for crop growth, analyzing data in the decision-making stage, and carrying out variable rate application using smart farm machinery.

Monitoring the area for crop growth conditions includes not only the health status of crops but also climatic information, environmental information, and growth information, and it is rapidly developing in both large-scale extensive agriculture,[6] as in the USA, and intensive

---

[5]Platform: In a dictionary sense, it means a flat place that is installed above the ground by the railway so that passengers can get on and off at the station. In the information age, the meaning is expanded to mean the hardware or software on which the computer system is based and the computer system that forms the basis on which the application program can be executed. In the first connection society like the Fourth Industrial Revolution, it is a place where people and information gather to create a new business. This is where the needs of suppliers and producers are exchanged, so that as more people participate, network effects occur and maximize value (e.g., Facebook, users communicate with their acquaintances, and providers connect with their ads).

[6]Extensive agriculture: used in relative terms with intensive agriculture. Use pesticides, fungicides, herbicides, etc. to reduce labor and capital inputs, and use machines in sowing, cultivating, and harvesting. Because yields are small per unit area, large crops are needed to earn revenue.

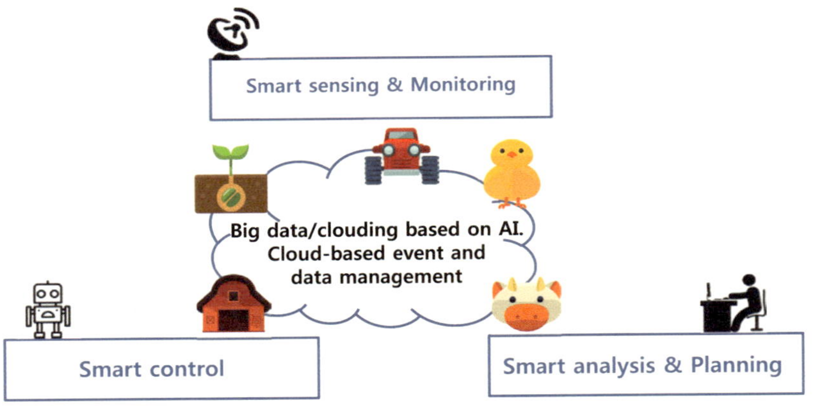

Figure 2. The Fourth Industrial Revolution and changes in agricultural production [2].

agriculture,[7] as in Korea. It is possible to maximize production volume and minimize the possibility of failure due to natural disasters, system errors, and other factors by acquiring data on growth, the weather, and agricultural equipment.

Analyzing data in the decision-making stage involves analyzing data from the monitoring stage and determining agricultural work required. In this stage, collected data is accumulated, processed, and analyzed as big data. Then, efficient and precise decisions about the data are made in a way that surpasses human intelligence, wisdom, and experience.

Furthermore, it is possible to collect environmental data on cultivation through an agricultural service platform using big data. The information can be used to evaluate market sale trends according to market preference analysis, and then the data (the cultivation environment, pest information, climate and weather information, soil fertility, topographical relevance, etc.) can be fed back to farmers to optimize production environments. In recent years, big data and artificial intelligence have been used to greatly expand the fields of genetic engineering with respect to agriculture and livestock. Within the premise of resolving laws, regulations, and ethical issues, it will be possible to cultivate edible crops and biota crops that grow in extreme climates or droughts. It will also be possible to transform animals' genes in order to make them more economical and suitable for local environments.

Variable Rate Application[8] using smart farm machinery is the third stage of this process. In the previous stage, the optimal decision was chosen for each location. In this stage, it is necessary to input the prescribed farm material suitable for the location. In extensive agriculture,

[7]Intensive agriculture: used in relative terms with extensive agriculture. Agriculture that maximizes economic efficiency by appropriately using tools and machines with the input of labor and capital

[8]Variable rate application (VRA): When there is a need to distribute the amount of farming material differently by location, the amount is changed mechanically. For example, when spraying herbicides, apply herbicides only in the presence of weeds, but increase the application rate in places where there are many weeds.

several tractors will be able to accomplish the same tasks (i.e., herbicide spraying) at different positions (i.e., variable rate application) by following certain intervals.

At night, when the farmer is asleep, a robot could be guided via FPS and electronic maps, enter the field, finish any necessary agricultural work, and return to the house before down. This dream will be a reality in the near future. It will be brought on by the Fourth Industrial Revolution.

3.2. Agricultural product distribution

Agricultural distribution is another field in which 4IR technologies will cause innovations. In each previous industrial revolution, the consumption pattern of agricultural products changed greatly. Prior to the First Industrial Revolution, 90% of the world's population was engaged in agriculture, so the distinction between producer and consumer was unclear. The First Industrial Revolution was an era of self-sufficiency in which the producers soon became the consumers. Raw materials were quickly consumed, and only very little raw materials were processed.

Through the Second Industrial Revolution, surplus products began to emerge so processing and storage technologies were developed. During this period, the agricultural production population shifted to manufacturing and service industries. The separation between rural producers and urban consumers became clear, thus increasing the necessity and importance of distribution.

During the Tertiary Industrial Revolution, the surplus product increased, and the central value of consumption moved from quantity to quality. Thanks to the increasing number of consumers, selective consumption has become more prevalent, and distribution functions have become more important.

The introduction of a customized agricultural product ordering system that takes into consideration the aging population and the expansion of single-person households in the agriculture and rural areas, including the control of shipment volume through the big data and the consumer's dietary style, suggests that the Fourth Industrial Revolution could revolutionize agricultural distribution.

Information such as the prices of agricultural production, crops, and distribution include the basic data necessary to manage supply and demand. By applying 4IR technology, comprehensive data, including agricultural production, climate information, population structure, and consumer data, are analyzed comprehensively. In this way, it is possible to produce customized products to optimize supply and demand autonomously. At the same time, the government can adjust timing and output in order to stabilize prices.

3.3. Agricultural consumption

During the Fourth Industrial Revolution, consumption is expected to be once again distinguished from previous revolutions. When consumer and producer information are linked in real time, it will be common to choose that best match both. 4IR technologies will also provide trade information through cyber and mobile production history and quality information. AI linked with big data will be able to stabilize transactions by connecting production information and transaction information.

For example, intelligent refrigerators will be able to automatically refresh its stocks in real time, based upon consumption. A refrigerator like this could also be linked to a system that manages family nutrition and health information. It could even cook food for family members based upon the nutritional needs of the individuals in the family.

Furthermore, 3D printing will allow people to be individually and creatively involved in the self-production of food, farm materials, agricultural machinery parts, and tools. Three-dimensional printers can even be used to make healthy functional foods for children and the elderly, including soft processed food that are easy to chew.

3.4. Influence on the rural environment and rural life

The Fourth Industrial Revolution will change production, distribution, and consumption as well as the rural environment and rural life.

At the same time, it will continue to develop agricultural systems by overcoming difficult problems that have yet to be solved by existing technology. It is expected that these techniques will be applied to actual farming sites, so they will require preparation and time for rooting.

4IR technologies can expand the agricultural industry diversely, from simple production-oriented agriculture to urban agriculture, healing agriculture, material agriculture, and industrial convergence. Examples of this include IoT, CPS[9], cloud-based agriculture experience and tourism materialization, aged farmer health information using wearable IoT, rehabilitation applied to animal and plant healing models, IoT and cloud, and urban agriculture using mobile technology. In addition, 4IR technologies are expected to find solutions to ongoing problems and malignant diseases that cannot be solved with existing technologies, such as animal odor, avian influenza, and foot-and-mouth disease. Above all, the 4IR will create new jobs by combining diverse technologies such as industrial convergence and hybrid technology. In addition, major changes will occur in risk management, bio-industrialization, and unmanned intelligence.

4. Preparing for the Fourth Industrial Revolution

In the era of the Fourth Industrial Revolution, new technologies and new businesses that cannot be defined by existing laws and systems will be developed. The positive regulation method of controlling gene expressions in gene therapy is currently illegal. In order to use

[9]Cyber-physical system (CPS): robots, medical devices, and real-time integration of software and environment in cyberspace.

positive regulation, businesses waiting to use new technologies and services will have to wait for laws to be enacted, which allow the use of positive regulation.

In order for the 4IR to be rooted in agriculture, it is necessary to promote the safety of agricultural work and rural life and to create a convenient environment for cyber technology and cloud infrastructures. This will prevent medical and cultural inconveniences in rural areas.

Wearable IoT and mobile devices are concrete methods in which we can implement agricultural work safety, cyber physics systems (CPS), remote medical, cyber cultural life, and aged farmer's life safety and health information big dataization.

4.1. Agricultural robots

A robot is a machine that moves independently, imitates humans, recognizes the external environment, and makes independent judgments about how to handle different situations. Agricultural robots will operate in every area of the agricultural process, including production, processing, distribution, and consumption. They will recognize the service environment and autonomously provide intelligent work or services. Agricultural robots can be defined as "intelligent agricultural production systems that can minimize human intervention, control themselves, and maximize efficiency." Traditional farming machines and unmanned aerial vehicles can be utilized by robots in the fields of agricultural product selection, automated distribution systems, facility horticulture, and automated livestock care.

Robot usage can be divided into three fields, depending on where they are used. These fields include open-field agriculture robots, facility agriculture robots, and livestock robots. These fields will aim to improve productivity through automation, unmanned farming, and the promotion of eco-friendly farming.

The global robot market[10] is expected to grow at a CAGR[11] of 17% from $ 71 billion in 2015 to $ 135.4 billion in 2019. The robotics market for agriculture and fisheries[12] is estimated to be $ 900 million in 2013 and is expected to increase rapidly to $ 19.1 billion by 2020. The target is expected to be a weed control and harvesting robot (see **Table 1**) [4].

4.2. Precision agriculture

Environmental problems continue to plague the Earth, yet the production of safe agricultural products is emerging. Interest in precision agriculture is increasing, in order to minimize environmental pollution and maximize the production of agricultural products. Scientists as well as those involved in agriculture are showing interest in this research. In

[10]The International Data Corporation (IDC) selected six technologies that were highly likely to grow: Internet of things (IoT), cognitive (recognition) systems, next-generation security, AR-VR, robot and 3D printing, and the market for each technology (2015.10.).
[11]Compound annual growth rate.
[12]International Federation of Robotics (IFR) 2014 Wintergreen Research report.

Division	2013	2014	2015	2016	2017	2018	2019	2020
Agricultural robot market	956	1386	2329	4634	8110	11,760	15,288	19,109
Growth rate (%)	34	45	68	99	75	45	30	25
Milking and livestock facilities	879	1203	1615	1918	2004	1735	1798	1611
High value-added crop	29	55	116	275	568	941	1376	1911
Cereal crops such as wheat, rice, corn, etc.	11	28	186	695	2109	2940	3669	4395
Grape pruning and harvesting	6	6	137	941	1272	1570	1413	969
Seedling management	14	42	116	292	616	1047	1682	2389
Grass management (lawn care)	14	43	140	371	811	1411	2410	3058
Unmanned aerial management	3	7	19	139	730	2117	3210	4777
(Source: World Robotics 2012. IFR).								

Table 1. The global robot market scale in agriculture and fisheries (Unit: million $).

fact, many are interested in precision agriculture because it does not belong to any one field; all fields contribute in a joint effort to solve the problems facing precision agriculture. Breakthroughs in agricultural machinery are of utmost importance; thus emphasis is being placed on engineering in the field. As the world's population continues to increase, there is an urgent need for an increase in food production. This need is hindered by industrial pollution and difficulty producing safe agricultural products due to harmful pesticides and fertilizers. Precision agriculture has emerged as a solution to this need, as it can increase the production of agricultural products while reducing the amount of harmful chemicals applied to the environment. Every crop field has different characteristics that can be measured in quality and quantity. Some examples of these characteristics include soil, nutrients, flow of irrigation water, and pest resistance. These differentiations of characteristics can all exist within a single crop field, so we have found that if we understand the different characteristics of each part of a field, and if site-specific processing is done for each location, the most profit from the least investment can be achieved. Therefore, precision agriculture follows the concept of variable rate agriculture. Yet it is prescription agriculture as well, as optimal profit is obtained based on past information. It has the ability to regulate future field conditions and yield through site-specific management. Precision agriculture is a concept that meets the needs of an advanced society that requires environmental preservation.

As shown in **Figure 3**, the concept of precision agriculture is one in which agriculture work is not actually made more precise, but instead the agricultural system as a whole moves from a statistical approach to a quantitative approach. Therefore, it is not an exaggeration to say that the scope of precision agriculture is the entire agricultural system. As a system of agriculture, three divisions of technology must be utilized in order to fully develop precision agriculture.

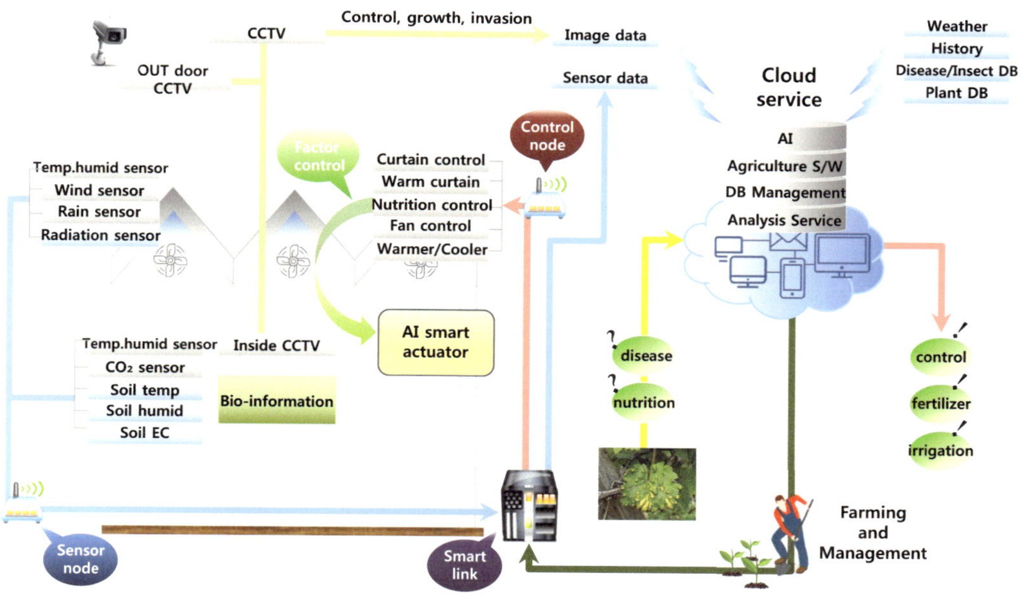

Figure 3. Crop production environment through biometrics and artificial intelligence (source: Convergence of agriculture R&D and Fourth Industrial Revolution, 28p) [5].

The first division is the acquisition of information related to the environment where crops will be grown, such as crop growth status, soil fertility, and climate by location. The means of obtaining such information is via sensors placed at each location, which can monitor different conditions including the yield of crops, moisture content of the soil, soil nutrients, moisture stress, and the occurrence of pests or weeds. These sensors do not collect information to be later analyzed in a laboratory; they are capable of instantly processing and storing information in real time.

The second division is the distribution of necessary, measured agricultural material into the crops. Based upon outcomes determined in decision-making and crop management, machinery will release seeds, nutrients, and chemicals to the crops.

The third division is the processing of computerized geographical information and databases along with the farmers' prescribed inputs in order to drive the control systems of various farm machineries. Even if the first two divisions are well developed, it is difficult to carry out precision agriculture if the third decision-making process is lacking (see **Figure 4**).

As one sector of agriculture changes and one farmer's agriculture becomes technologically advanced in this way, it does not mean that precision agriculture has been established. That is, precision agriculture does not change farm by farm. Precision agriculture is not a word referring to a single technology, but an overall concept of new changes in agriculture.

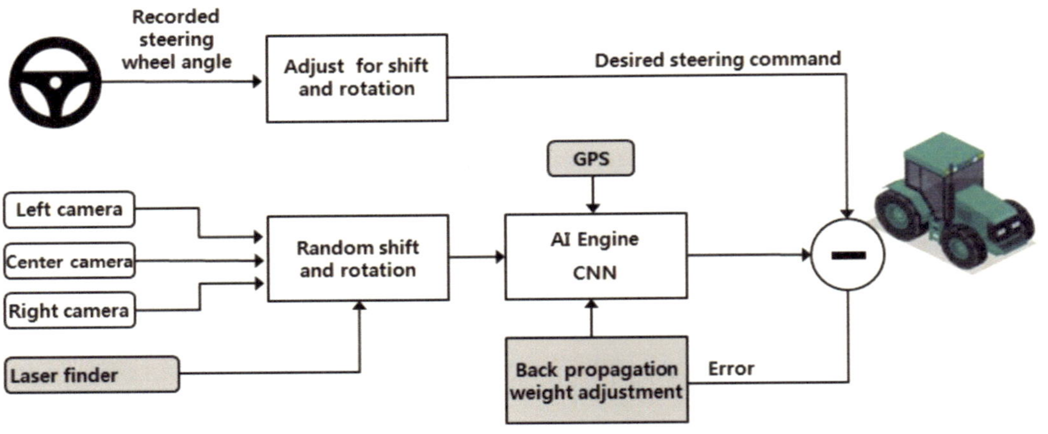

Figure 4. Artificial intelligence autonomous driving and unmanned agricultural farm machinery (source: Convergence of agriculture R&D and Fourth Industrial Revolution, 37p) [5].

5. Conclusion

Just as the first, second, and third industrial revolutions did, the emergence of new technologies achieved via revolution always begins with the destruction of an existing order. Breaking the existing order creates a gap in which opportunities can emerge. The Fourth Industrial Revolution technology presents a chance to increase agricultural competitiveness and an opportunity to overcome the structural weaknesses of our current agricultural system and the limits of intensive agriculture. There are three steps that we must take in order to lead the change.

First, we must analyze the impact of the 4IR on our agricultural ecosystem. It is necessary to analyze the impacts on all fronts of agriculture, the effects on rural and agricultural life, and the effects on agricultural structure and work.

Second, we must consider data management and its effects. In the future, data will be a resource, and data quality will be competitive. Data should be standardized so that quality agricultural data can be continuously produced and managed.

Third, we must facilitate the construction of an infrastructure that supports technology-based agriculture. The fifth-generation (5G) communication network, the Internet network infrastructure, and the Cloud Service System must maintain support for these technologies in order to allow them to integrate easily into the agricultural industry.

If research and development supports the fusion between heterogeneous technologies and heterogeneous industries and the agricultural industrial ecosystem allows creative talents to freely exercise their capabilities, the Fourth Industrial Revolution can occur. In this way, agricultural technology will grow to new heights and leap to new opportunities.

Author details

Jehoon Sung

Address all correspondence to: jhsung@korea.kr

Rural Development Administration, Republic of Korea

References

[1] [Internet]. Available from: www.wikipedia.org [Accessed: July 13, 2017]

[2] Lee JR. The Fourth Industrial Revolution and Future Agriculture. Rep. of Korea: Science and Technology Policy Institute; 2017; 200 ed

[3] Lee HJ. The Fourth Industrial Revolution and the Future of Agriculture. Smart Farm and Shared Economy. Rep. of Korea: Science and Technology Policy Institute; 2017; 200 ed

[4] The Fourth Industrial Revolution and Agriculture. Rep. of Korea: Korea Institute of Planning and Evaluation for Technology in Food, Agriculture and Forestry; 2016

[5] Lee YB, editor. Convergence of agriculture R&D and 4th Industrial Revolution. 1st ed. Jeonju, Rep. of Korea: SY Ra; 2017. 117 p

Chapter 2

The German Vision of Industry 4.0 Applied in Organic Farming

F. J. Knoll and V. Czymmek

Additional information is available at the end of the chapter

http://dx.doi.org/10.5772/intechopen.72708

Abstract

The first industrial revolution was the invention of the steam engine. With the advent of conveyor belts and electricity, the second industrial revolution arose. After the third revolution, the automation, the fourth industrial revolution takes place with the complete networking of all machines, workers, consumers, and products. In Germany, this is called Industry 4.0. Increasing digitization makes it possible to collect, store, analyze, and communicate large amounts of data. By digitizing farms, a network of different sensors can analyze the nutrient content and the soil texture in real time. This information can be evaluated and the plant distribution can be managed across all networked farms. This leads to the right field being used for the right plant at the right time. Real-time data processing makes it possible to monitor and control the nutrient intake over the entire growth period. This allows the field to specifically ask for water or the right fertilizer for its plants. This saves resources and protects the environment. All the prepared information can give the farmer an exact status about his products and fields via an interface. This horizontal networking within the farm and the vertical networking across different farms can lead to increased efficiency and cheaper products. The use of robots can create a fully automatic farm. For this undertaking, it is necessary to process the complex information of a farm with a self-learning system. At the Westcoast University of Applied Science, for example, a robot is being researched to automatically remove the weeds. The prototype of the robot that moves fully autonomously across the field classifies the plants and destroys the weeds.

Keywords: CNN, AI, robot, industry 4.0. organic farming, plant detection

1. Introductions

For the classification of the plants, a self-learning system is used, which learns to recognize different plants [1–3]. These artificial neural networks are part of the most recent artificial intelligence research and are used for a variety of different applications, such as autonomous driving [4, 5], for airline passenger profiling [6], or information processing on the Internet. The application makes it possible to minimize the use of herbicides or to avoid them completely by a nonchemical destruction. This type of environment friendly agriculture would only be possible with many workers who weed the fields by hand. However, the resulting costs do not allow the production of bio-economic products for the mass market. With an all connected system, the farmer is always up-to-date. For example, a drone could tell the robots which field needs to be processed at this point. Other robots could take sowing and harvesting. As a result of the progressing hardware minimization at Westcoast University [7], these agents could be placed directly in small drones in the near future, and a swarm could handle the fields. In this chapter, a vision of the autonomous farm is described. The basic vision and necessity due to the increasing lack of assistants will be explained in section 1. In section 2, the project of the former co-pilot of the Westcoast University of Applied Sciences is presented as an example for the digitization of a work step on the farm. Section 3 presents various sensors and algorithms that are already used for plant selection. Section 4 describes the use of a convolution neural networks, which corresponds to the mathematical reproduction of a brain, for the information processing of a genetic robot. Section 5 explains the environmental friendly destruction of the robot. In the penultimate section 6, future objectives, such as the swarming intensity of different drones, are described. In the last chapter, the advantages and dangers of the digital revolution are shown.

2. What Industry 4.0 means in agriculture

The German version of the Industry 4.0 is made up of the following components:

1. A complete network of machines, sensors, devices, people, and products via the Internet.
2. A virtual image of the real world is created via sensor data, thus building up an extended information system.
3. Assistance systems help people through visualized interfaces to preprocess and filter the information they have acquired, so that faster decisions can be made. In addition, physical, monotonous, strenuous, and dangerous work should be physically relieved.
4. An information processing system should make all decisions and autonomously control the entire process. It should take decisions independently. Only in the case of problems or target conflicts, the powers should be transferred to a higher authority [8].

If the concept of Industry 4.0 should be transferred to agriculture, a central learning system will take over all the decisions, processes, and problems of the farm. This means that different daemon units measure the fields via sensors, for example, the ground conditions. The moisture or nutrient content is then sent to a database. This creates a virtual and real-time image of the farm. In addition, global information could be obtained via the Internet. The learning system will process past and up-to-date information in real time to create optimal growth conditions for each plant. The system can continually focus on each individual plant. For example, it could determine the best position for each crop on a field. It is conceivable that different plant species are sown on the same field in order to optimally exploit the soil texture. It could access global weather data to calculate the optimal time of the growth cycle. Automated machines could travel independently to the fields and process them. Drones could monitor everything and determine further information about the growth behavior or the environment of the plants. The plants could be sown to new, more precise patterns, in order to better organize the detection and nutrient distribution. Nutrients could be targeted to individual plants, so that each grows optimally. The system could find through historical records, the connection between soil and plant species, which an employee cannot capture. The customer could receive a biography about the whole life of the plant with all influences from the seed to the fruit. All these networked small self-sufficient systems would take over the full farm automatically and manage it.

Furthermore, with VR glasses like the Hololens of Microsoft, human coworkers could get important information about individual plants.

The system could independently process global economic data and develop new economic sectors through the Internet.

Further, the system could develop its own improved devices to optimize the work. Since this system can work 7 days per week and 24 hours a day without getting exhausted, it can deal much more thoroughly with cultivation than a farmer in mass production. Food production for the current human lifestyle has become increasingly polluting. In conventional cultivation, more harmful herbicides and pesticides are used to increase efficiency. It is true that a very efficient pest control is carried out and thus the yield is increased, but this use of chemical also has some side effects. One of the best-known consequences is the increasing pollution of groundwater and drinking water or the death of bees. These two problems alone could put mankind faced with unprecedented challenges. In addition, the herbicides and pesticides pollute the food itself. Therefore, some farmers are trying to establish a more biologically neutral agriculture. Like the farm is Westhof in northern Germany. This is one of the largest farms for organic products in Germany. Due to its environmentally friendly use of the fields, it is prohibited by law to use herbicides in Germany. For example, the farm employs every season a group of workers to free approximately 170 hectares of carrots from weeds. The weeds must be weeded, because the plants compete for nutrients and thus reduce the yield. The Westhof provides more than 170,000 € per year for weeding. Getting workers for these jobs is becoming more and more difficult due to the growing standard of living. Although there are already technologies that have been automated in agriculture, for example, self-propelled tractors or various attachments, at the present time, weeding can only be done by hand. Especially when it concerns plants like the carrot. For this reason, the West Coast University of Applied Sciences in Germany wants to develop a fully automatic weeding robot. This research project is called high-precision weed recognition in organic farming. The autonomously working robot is supposed to take over the weeding day and night in any weather conditions. The project is already in an advanced phase. This development is an automation of a work step but it can also be included in point 3 of the definition list under assistant systems. With its extensive facets, it also provides an insight into the difficulties of the digitization of a farm. Therefore, this project should be used in the following chapters as an example for Industry 4.0 in farming.

3. Project highly accurate weed detection sensors

The first step toward autonomous eliminating weed is to find out which type of sensors is the most efficient for classification and detection. It is also important to record when the ideal time for weeding is. The co-worker weeds at a late growth period. At this point, the plants already overlap each other. It is therefore not possible to see the weeds under a plant without mechanical effort. For the workers, this is the ideal time, because the advanced growth allows to differentiate individual plants by different characteristics. The robot weeds at an earlier date when the plant and the weed look very similar. This is shown in **Figure 1**. While this allows an unrestricted view of the plants, it is more difficult to distinguish between them because of their less pronounced characteristics.

Figure 2 shows the robot. The BoniRob has four freely adjustable arms on which the wheels are suspended. Each wheel is driven by its own electric motor and the arm adjustment allows any axis spacing to be set. In addition, the robot has a mounting slot in the center, in which various measuring instruments can be installed [2].

For reproducible results, the sensors were sealed off from ambient light. So the influence of the sun on the measurement results can be prevented. A LED-lighting, consisting of six high power LED's mounted on a profile, was implemented to illuminate the plants properly. The LED's have a color temperature of 6000 K, which is set at daylight. The lamps are arranged in a circle at a certain angle, so that they have a fixed focus point on the field. By this arrangement, any shade can be prevented and a homogeneous illumination can be achieved [2].

With this configuration, the robot runs autonomously at a speed of 20 mm/s over the field. This speed allows an uninterrupted recording of the field at one image per second, with an overlap of the respective sensor data of approximately 20%. The area on which the plants grow on the wall has an average width of about 15 cm. In order to determine a suitable sensor for recording the plants, the following were investigated: [2]

- Bispectral JAI camera.

This camera has a resolution of 1296 × 966 pixels. In addition, it allows the acquisition of coverage-equivalent and simultaneous images from the color range and near infrared range. Two bispectral cameras were always examined at the same time at different angles [2].

Figure 1. Recorded plants by the sensor. Weed encircled red [2].

Figure 2. Bonirob on the field of the Westhof Bio in Germany.

- Nikon camera D5300

The Nikon D5300 camera has a resolution of 6000 × 4000 pixels. These images do not have an infrared color channel [2].

- Kinect II

The Kinect II offers the possibility to record a color image with the size of 1920 × 1080 pixels. The time of flight technology allows a depth of 512 × 424 pixels with 16 bits. An infrared image of 512 × 424 pixels and 11 bits can be read out via the same sensor. However, the depth image and the color image are not coincident and are recorded over two different perspectives [2].

- CamCube 3

The CamCube uses the time of flight technology to create depth images and has a resolution of 204 × 204 pixels. Unlike the Kinect, it is possible to use own lenses. As a result, optimal image sections can be seen. A congruent gray image is displayed next to the depth-image [2].

- LMI Gocator 2350

The LMI Gocator 2350 is a laser scanner with a resolution in the z direction of 0.019–0.060 mm and a resolution in the x direction of 0.150–0.300 mm. The field of view (FOV) is 158–365 mm [2].

These sensors were evaluated between 22 July 2015 and 14 August 2015 every day under different weather conditions over 1000 meters of rootwalls. By means of these data, it was possible to determine which sensors, configuration, and which stage of growth are best suited for the classification of root plants and weeds [2].

The uneven background and the low resolution of the CamCube 3 prevent a classification between a plant and dirt. The Kinect II has a similar problem. An 3D image of the Kinect is shown in **Figure 3**. Although it has a much higher resolution and a RGB camera, this sensor is intended for use in a living room as a game interface and therefore has a small focal length. As a result, over 2/3 of the resolution is lost to unimportant areas adjacent to the root rows. The focal point auf the RGB camera is not suitable and produces blurred images due to the distance to the object [2].

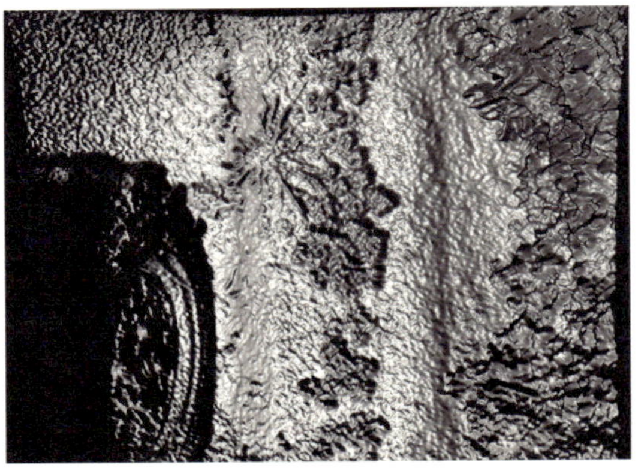

Figure 3. Depth map of the Kinect II [2].

Figure 4 shows a picture of the LMI Gocator 2350. In the picture, different plants can be seen clearly. But the sensor cannot correctly detect the plants in the lower half of the image. Among other things, this is due to the fine structure of the plants and the measurement resolution of the sensor. In addition, there are difficulties to use the LMI Gocator 2350 for plants that are currently in the bud stage. These plants still have a very small area and are located very close to the ground, which means they get lost in the overall information. If the ground is not flat, then it is almost impossible to filter the small structures [2].

The best results are provided by the JAI camera and the Nikon camera. In contrast to the Nikon camera, the JAI has an infrared channel in addition to the RGB channel. The RGB sensor and the IR sensor use the same optics. Thus, it is not possible to focus both images at the same time. However, with this 4-channel image as described in the next chapter, the plants can be extracted from the background by a vegetation index method. The Nikon camera has a high-resolution color camera. Later, instead of the Nikon camera, a 42 megapixel camera with 7716 × 5364 pixels was used. As described in the following chapter, the project has succeeded in creating a robust and real-time segmentation algorithm for RGB images. The used RGB camera is in contrast to the JAI a standard camera and thus easier to interchange, so the choice came down to RGB image recordings. A detailed evaluation of the sensors, as well as a description of the measurement setup, can be found in the paper [2].

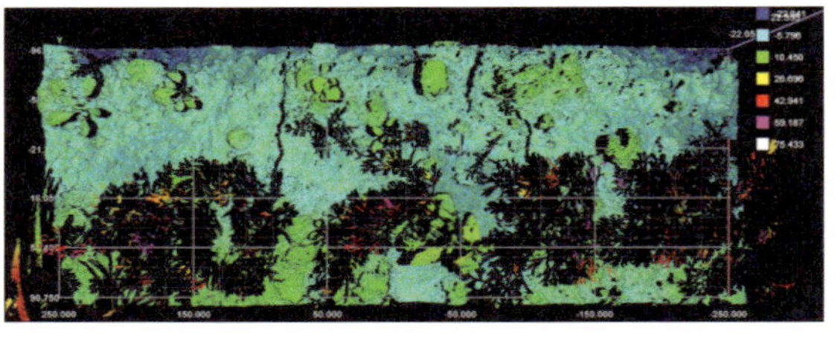

Figure 4. Recording of the field with the LMI Gocator 2350 [2].

4. Segmentation

The next step for the classification is to separate the plants from the background. The separation of the plants from the background allows the classifier to examine only the important plant pixels in the image. This approach therefore saves time for the classification of a whole picture. The state-of-the-art method used a bispectral camera. Beside an RGB image, this camera also delivers an image from the near infrared region. Both images are congruent because they are taken through the same lens. The plants can be extracted from the background by simply subtracting the red color channel and the infrared channel and then using a global threshold. The state-of-the-art vegetation index determination method works, because the chlorophyll content of living plants causes a greater increase of reflection in the near infrared spectrum than other objects. The big advantage of this method is that dead materials like stones, sticks, or wilted leaves are filtered out. Hence, the extracted plants can then be easily selected and classified. The harsh environment on a field requires robust sensor systems. The mechanical setup of bispectral cameras is more complex compared to standard RGB-color cameras. Hence, standard RGB cameras are more suitable for organic farming applications. Furthermore, they are cheaper and better available, as a wide range of RGB cameras exist on the market. Hence, not only the robustness is better but also the total system costs decrease. This is an important feature as the acceptance of the organic farmer for such systems are only given if the price is much lower than the costs for manual weeding. Another way to extract the plant from the background is to process the HSV-Color-Space. The proposed workflow is shown in **Figure 5**. In the first step, the captured RGB image is transformed into the HSV-Color-Space [3].

The HSV-Color-Space uses the colors with the three vectors H for Hue, S for Saturation and V for Value which stands for the brightness. The exact working method is presented in [9] and the result of the algorithm can be seen in **Figure 6** [3].

For a better quantitative result, the ground truth was manually generated. The plants were colored by hand in multiple images. It should be obvious that this method is a subjective one. With the ground truth, it was possible to create a statistical number of over 250 plants. Afterwards, the dice score [10] was calculated

$$DS = 2 \cdot \frac{|X \cap Y|}{|X| + |Y|}. \tag{1}$$

The set X is the ground truth and the set Y is the mask of the vegetation index method used (state-of-the-art, proposed HSV, or RGB algorithm). It should be obvious that if the set X and Y are identical, the calculated mask and the ground truth are equal, hence DS = 1. The dice score and its standard deviation can be looked in **Table 1**. As it can be seen in **Table 1**, the proposed methods are better as the state-of-the-art vegetation index method described in [11].

A high-resolution camera (6000 × 4000 pixels) was used in a further series of measurements. Again, all the plants could be cut out with the two presented methods without any problems and with a high level of detail, only the state-of-the-art vegetation index method does not work

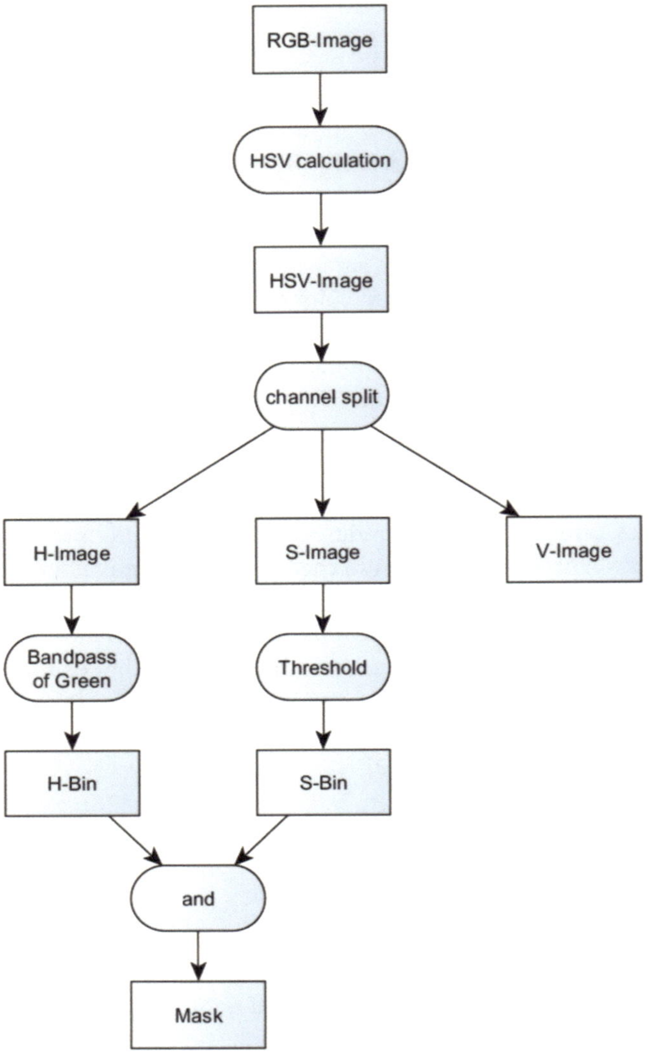

Figure 5. Workflow of the first alternative vegetation index determination [3].

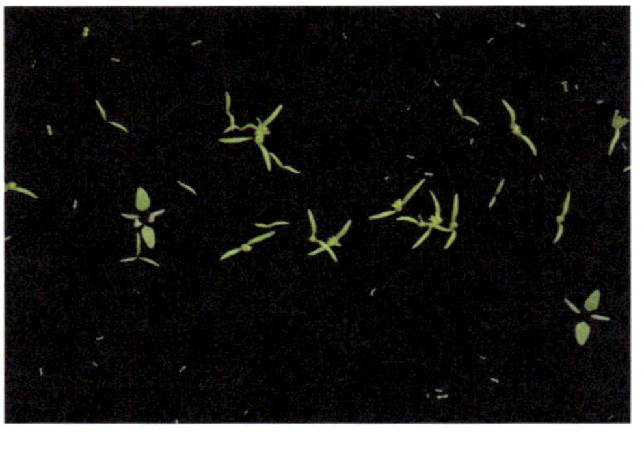

Figure 6. Result of the segmentation.

Methods	⟨DS⟩	Standard deviation
State-of-the-art	0.812	0.063
HSV	0.962	0.014
RGB	0.943	0.018

Table 1. Dice score and standard deviation of the state-of-the-art and the proposed vegetation index methods [9].

here, because there is no IR image present. With the higher level of details, the classification task becomes easier as more features are available. This will be a significant advantage of the two proposed methods [9].

A more thorough investigation of the algorithm as well as a mathematical derivation and optimization can be found in [3, 9]. A further RGB algorithm is also presented here.

5. Classification

After filtering the background, the plants need to be classified. Therefore, the images of the extracted plants have to be classified manually to create a learning mask. These can be used to train a classifier. There are different classification algorithms, for example, the state-of-the-art random forest classification procedure [11]. One problem of the random forest classifier is to find and calculate suitable features. Another method which is used in this paper, is a convolution neural network classifier, which has to be trained as well and is used in two different ways as described in the following paragraphs [9].

The first process shall evaluate each pixel individually and specify if it is a carrot plant or weed [9].

5.1. Setup of the proposed convolution neural network

The convolution neural network for classifying the pixels manner consists of 11 layers and is using a RGB image of 101 × 101 pixels as input information. It is the same structure as in **Figure 7**, but the output logic is able to classify three classes.

The convolution neural network consists of:

- Input layer, size of 3 × 101 × 101
- Convolution layer, size of 16 × 5 × 5 (Leaky rectify)
- Convolution layer, size of 32 × 7 × 7 (Leaky rectify)
- Pooling layer, size of 2 × 2: 2
- Convolution layer, size of 32 × 5 × 5 (Leaky rectify)
- Convolution layer, size of 64 × 7 × 7 (Leaky rectify)
- Pooling layer, size of 2 × 2: 2

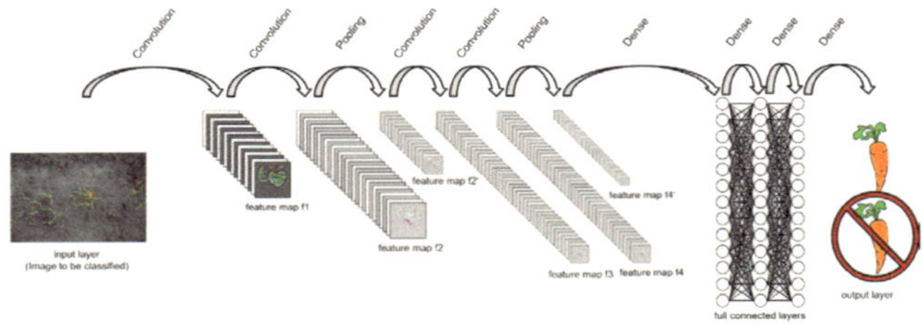

Figure 7. Example CNN for classifying carrot plants in organic farming [9].

- Dense layer, size of 64 (Tanh)
- Dense layer, size of 64 (Tanh)
- Dense layer, size of 64 (Tanh)
- Output layer (Dense layer), size of 3 (Softmax)

The output layer consists of three neurons. As a result, three classes can be classified like carrots, weed, and background [9].

5.2. Training from the convolution neural network

This proposed classifier was trained using a CUDA on a GTX Titan with 6 GB memory. It was used a batch size of 500 images. After a learning period of about 3 weeks and 950 million periods, the convolution neural network has been tested [9].

To evaluate the proposed CNN algorithms a huge image database has been build up. The images in the database were taken with a robot, which is shown in **Figure 8**. The robot named BoniRob drove with constantly 200 mm/sec over the field and took an image every second. These measurements were repeated daily for several weeks under different weather conditions. The images were taken with the Nikon D5300. Seven hundred plants got evaluated. If the majority of the pixels are identified as a plant, it belongs to this class. The resulting

Figure 8. Bonirob on the field of the Westhof Bio in Germany.

classified images of the plants were compared manually to the input images to examine the error rate which was less than one percent at 700 plants. An error means the classification of a weed as a carrot. A carrot was never classified as a weed. This is important for the farmers [9].

Figure 9 represents the weed in red and carrots in green. This image was created pixel wise but only every tenth pixel to save time. The white circle shows that the CNN classified a carrot at a weed plant. However, we classified this manually as only a weed plant, which was wrong. Hence, the network corrected us [9].

A better analysis of the accuracy should be done again via the dice score Eq. (1). [10]. Several convolutional neural networks have been designed to perform the classification task. The exact operation is presented in [9]. To train the CNN, other images were used than for testing it. These were unknown test data. The images got first classified by hand, so that a ground truth can be compared. An original image, a ground truth image, and a classified image by the proposed pixelwise CNN from the same area can be seen in **Figure 10**.

The classified image with the pixelwise CNN in **Figure 10** is smaller, as the origin and the ground truth image. The three in [9] proposed Convolution Neural Networks (pixelwise-, skelton- and area classification) are compared to each other and the results are shown in **Table 2**. It should be noted that the calculation of the dice score was based on only the pixels that were not classified as background in the ground truth image and the images of the presented algorithms. This is because in the marginal area of a plant, the threshold value does not always calculated exact the same edge lines and thus a misleading error would be calculated.

In **Table 2**, it can be seen that the number of classifications of the carrots and the weeds are different. For example, the weed plants are less often seen in **Figure 10** as the carrots. While the carrots are relatively homogenous in size, most weeds are relatively small. Therefore, its number of pixels is less than that of carrots. However, it was made sure that a relatively equal number of small and large weeds were used in learning. This could prevent the convolution neural network from triggering only to the size [9].

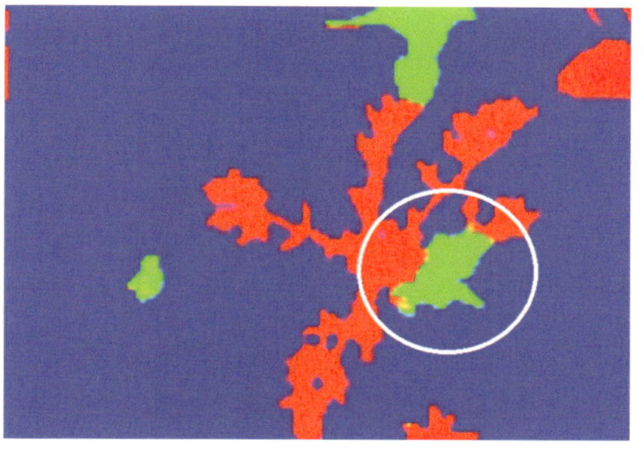

Figure 9. First result of the CNN [9].

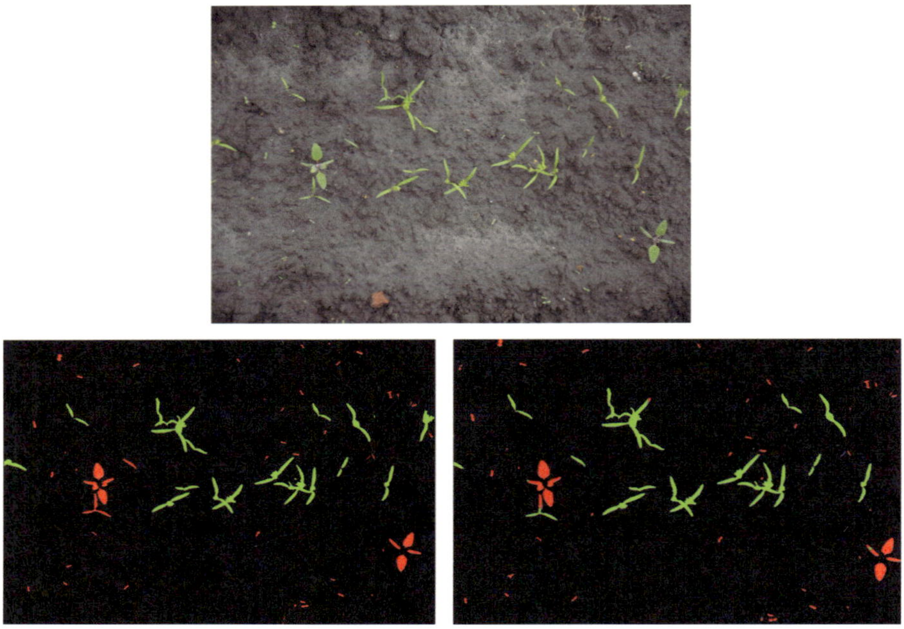

Figure 10. On top, the original image, bottom left side, the ground truth classified by hand image and on the bottom right side, the classified image by the proposed pixelwise convolution neural network.

It can be seen from the **Table 2** that the accuracy for the recognition of a carrot pixel is over 98% for all the presented methods with a standard deviation of approximately 2%. The presented area classifier even reaches an accuracy of 99.6% with a standard deviation of only 0.5%. The weeds, on the other hand, have an accuracy of 88.1% and the skeleton of 89%. This means that around 11% of all weed pixels are classified as carrot plants. In the area classification, the accuracy is over 96% for the weed detection and thus almost as well as for the carrots. The default deviation is only 4.3%. The area classifier shows the best results. A complete statistical evaluation can be considered in [9].

Methods		⟨*DS*⟩	**Standard deviation**	**Number of image**	**Number of classifications**
Pixelwise	Carrot	0.986	0.020	45	1,339,190
	Weed	0.881	0.102	45	188,869
Skeleton	Carrot	0.983	0.024	44	164,781
	Weed	0.889	0.077	44	25,148
Area	Carrot	0.996	0.005	25	808,069
	Weed	0.968	0.043	25	120,840

Table 2. Dice score and its standard deviation of the three proposed convolution neural networks.

6. Destruction

The classification results are then transferred to a destruction unit. This destruction unit should remove the weeds environmentally friendly. How exactly this destruction unit will work is examined. There are many different considerations. The methods must have a number of properties. The most important feature is the environmental compatibility that is required by organic farming. Therefore, the use of herbicides is forbidden. Furthermore, the method must be very robust. This means that it has to work several thousand kilometers of carrot rows without great maintenance. Therefore, for example, a method in which the weed is plucked would be too complicated. Looking at the individual steps of the plucking process, it is noticeable that a variety of sensory and motor skills are required to pull a plant out of the soil and not tear it down. Therefore, mechanically simple approaches result. In the following some methods are listed:

- Stamping

While stamping, the plant should be removed by pressing it into the soil. It is not completely eliminated, but the crop has an advantage in the distribution of nutrients.

- Hot bolt

Similar to stamping, a hot tip should be used here in order to burn the bicarbonate in the growth center.

- Mill

A small milling cutter turns the weed from the ground.

- Laser

The plant gets burned by a laser. Problems could be the required power, and the lens system that might be stained in the harsh environment.

- Maser

Similar to the laser, bundled microwave beams could be used. The advantage would be that this radiation would also get under the earth. It is questionable whether microwave radiation is accepted in ecological agriculture. In addition, the question arises if all the necessary life is removed from the soil.

- Electricity

An induction of current could burn the plant, or cause a chemical process in the plant, so that it enters. First experiments with a Tesla coil have been initiated. Targeting of the Tesla beam could be by means of ionized air.

- Matter beam

A water cutter/sandblaster could destroy the weed. A problem would be to carry the materials.

Another important feature of the destruction is the speed. In addition, as explained in the next section, further miniaturization is needed. Therefore, it would be desirable if the destruction unit also has a small space and energy requirement.

7. Future objectives

One of the most difficult problems in the use of artificial neural networks is the computational capacity. Although large search engine companies own specially developed hardware to provide the necessary computing power, for the conventional user remains only the state of the art method which is using a graphic processing unit (GPU) as a computational basis. Although these processors are well suited for large matrix computations, they need massive energy. Therefore, a new processor on the basis of a FPGA was developed. The power should be minimized because the use of a FPGA is only a fraction of a GPU application and should therefore be well suited for independent calculations.

The increasing computing power in recent years allows more and more complex algorithms for the implementation of adaptive artificial intelligence. The developed hardware processor consists of three main components: the control unit, the data path and the data manager. The architecture of the proposed hardware processor is illustrated in **Figure 11** on page before. The complete configuration of the CPU is shown in [7].

The control unit is responsible for determining what operation the processor should just execute. In the program memory, each individual step is deposited. As shown in **Figure 11**, the core of the control unit is the instruction register (IR). In the IR, the individual steps for each task are written. With each step, a program line is loaded from the program memory [7].

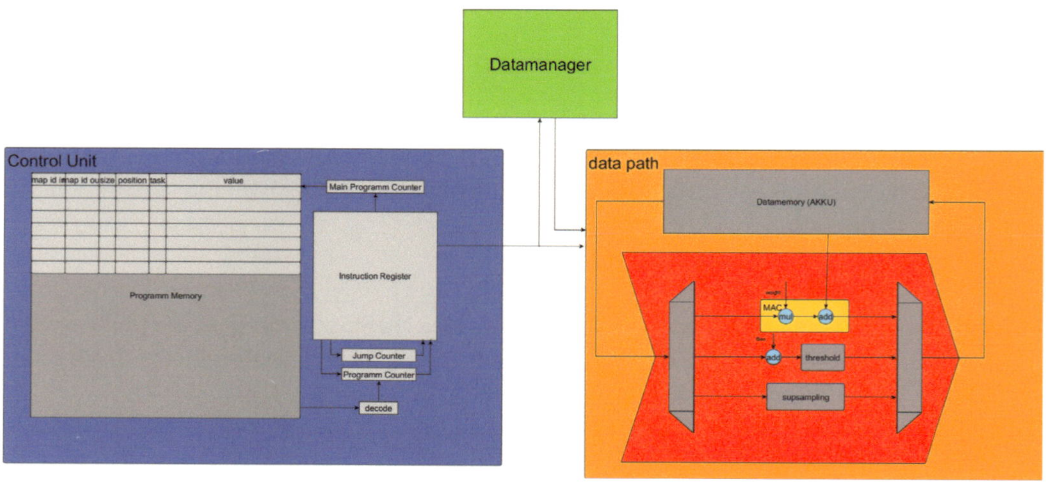

Figure 11. The architecture of the proposed CPU for calculation of convolution neural networks [7].

The data manager contains the memory management unit (MMU). It saves the saving address and its information. These are, for example, the starting addresses of the different feature maps [7].

The data path (see **Figure 11**). The accumulator (AKKU) contains all the data that has to be processed. The data manager knows where to find the information in the AKKU. The arithmetic logic unit (ALU) is responsible for the calculations. **Figure 11** shows the multiply and accumulation (MAC)-, the threshold-, and the subsampling-unit. The MAC unit is used for example for the convolution and dense layer [7].

In the future, the CPU will be further developed so that it can be used for plant recognition. The big problem now is that the used FPGA has too little memory to process large networks as needed for plant classification. Therefore, a more suitable FPGA should be used [7].

In paper [7], there is more in depth information about the developed CPU.

In upcoming projects, the existing robot will be optimized by a more cost-effective and more efficient design. Through miniaturization in industry and research, a large number of new technologies have been developed over the past decades. As mentioned in the beginning, it is always difficult to find coworkers for weeding in the field because of the substandard working conditions. Several workers are pulled by a tractor in a reck across the field and weed, the weeds laying faced down to the field. In order to protect the employees from harmful work, a transport device operated by electric motors and a solar panel on the roof was developed. This drives the workers without pollutants and noiselessly over the field.

The robot presented here was the first step toward the automation of agriculture. However, the current BoniRob in its construction with a weight of about one ton and the size of a passenger car is very bulky. After a long trial phase in the field, several disadvantages have arisen. Due to its dimensions, the transport is made more difficult and maneuvering in bad weather conditions is not always possible due to the soil condition. Another disadvantage is the high price of approximately 240.000 €.

To optimize these criteria, the Bonirob will be miniaturized in the future. First, the principle of the electric rack should be transferred to the BoniRob. An extermination unit is installed at the position of the coworker. A central arithmetic system classifies the recorded images and passes the position of the weeds to the destruction unit. Several batteries and a solar panel supply these units. A system allows to run autonomously over the field.

A possible following miniaturization in the future is the use of a drone swarm. The first experiments can be seen on **Figure 12**. Cameras at the bottom of the drones take pictures of plants. In order to save energy, the images and the GPS coordinates of the images are sent wirelessly to a central server, where the energy- and performance-intensive classification and detection are performed. Subsequently, the data is sent back to the drone and the weed will be destroyed by an extermination unit.

By the complete networking of the sensors in the Industry 4.0, the robots would receive, for example, weather data and decide when the best time to process the fields and when they have to fly back to their station to protect themselves or recharge the batteries.

Figure 12. First drones experiments on the field of the Westhof Bio in Germany.

8. Risks of automation

Through the complete automation, human workers would not be necessary anymore. It is possible that knowledge gets lost. It seems not to be required that someone still needs the extensive knowledge of a farmer. Yet no one could know how the weeds differ from the crop as this is done by the machines. If, for example, a solar storm happens and destroys all electronics, we would be sent back to the Stone Age. At the moment, we did not have the knowledge needed to manage fields. This example applies, of course, to any other branch of industry where artificial intelligence takes over. Therefore, it is imperative to pass on this knowledge, despite the seemingly nonexistent necessity. Again, the autonomous systems could help us. Through special interfaces like the hololence, the system could inform us about its fieldwork experience and explain new insights. Humanity would also be involved in the future and ready to take over the work. Another disadvantage is the high capital needed to invest in automation. The complete connected automation system mentioned in this chapter would cost nowadays millions of euros. With the increasing amount of connected systems and their communication, manipulations and data safety are other aspects to consider.

9. Conclusion

The robot is only the first step in the development of a fully agricultural farm. Since it is very large and expensive, drones or small beetle-like robots are already considered in future visions. These small autonomous systems could support each other by means of swarm intelligence and process the fields. Not only weeding should be considered, but also the targeted removal of pests. This would enable a completely pesticide-free and herbicide-free agriculture. The robot presented makes it possible to distinguish the plants in small stages of growth with a higher precision than humans can. He also processes the first plant with the same concentration as the last plant. This project represents a step toward the mass-efficient organic agriculture in which the environment is spared and more natural food can be produced. Through these robots and the arrival of industry 4.0 in agriculture, production prices could fall and high-quality foodstuffs could be produced for the whole world.

Acknowledgements

This work was supported by the Federal State of Schleswig-Holstein, Germany. The authors are grateful for the financial support.

Biographies

F. J. Knoll. Coast University of Applied Sciences (FHW), Heide, Germany, in 2009. After attending a physic study at Uni Hamburg, he made his M.A. degree at the Westcoast University of Applied Sciences (FHW), Heide, Germany. He is currently working on his doctoral thesis. His research areas include Convolution Neural Networks, real-time 2D/3D image processing, embedded systems design, and algorithm design for FPGA implementation.

Vitali Czymmek received the M.Sc. degree from the West Coast University of Applied Science, Heide, Germany, in 2017. He joined the project High-Precision Weed Recognition as a working student in 2015 and works now as a research associate thesis. His main areas of research interest are digital imaging processing, industrial automation solutions, and their control systems.

Author details

F. J. Knoll* and V. Czymmek

*Address all correspondence to: knoll@fh-westkueste.de

Westcoast University of Applied Sciences, Heide, Germany

References

[1] Federal Ministry for Economic Affairs and Energy. 2017. URL: http://www.plattform-i40.de/I40/Navigation/EN/Industrie40/WhatIsIndustrie40/what-is-industrie40.html

[2] Knoll F, Holtorf T, Hußmann S. Investigation of different sensor systems to classify plant and weed in organic farming applications. SAI Computing Conference. 2016:343-348

[3] Knoll F, Holtorf T, Hußmann S. Vegetation index determination method based on color room processing for weed control applications in organic farming. In: I2MTC 2016, Proceedings of the 33th International Conference on Instrumentation and Measurement Technology., IEEE Instrumentation and Measurement Society; 2016, pp. 1024-1029

[4] Jung S, Lee U, Jung J, Shim D. Real-time traffic sign recognition system with deep convolutional neural network, Ubiquitous Robots and Ambient Intelligence (URAI), 2016. In: 13th International Conference on, IEEE Xplore 2016

[5] Li J, Mei X, Prokhorov D. Deep neural network for structural prediction and lane detection in traffic scene, neural networks and learning systems. In: IEEE Xplore 2016

[6] Zheng YJ, Sheng WG, Sun XM, Chen SY. Airline passenger profiling based on fuzzy deep machine learning, neural networks and learning systems. In: IEEE Xplore 2016

[7] Knoll F et al. CPU architecture for a fast and energy-saving calculation of convolution neural networks. In: SPIE Digital Optical Technologies International Symposium 2017, IEEE Xplore 2017

[8] Hermann M, Pentek T, Otto B. Design principles for Industrie 4.0 scenarios. In: 2016 49th Hawaii International Conference on System Sciences (HICSS). 1 January 2016, S. 3928-3937, DOI: 10.1109/HICSS.2016.488 ieee.org, 22 August 2016

[9] Knoll F et al. Vision Based Measurement System for organic farming applications using Deep Learning. IEEE Trans. on Instrumentation and Measurement, submitted for publication 2017

[10] Dice LR. Measures of the amount of ecologic association between species. Ecology. 1945;**26**(3):297-302

[11] Haug S et al. Plant classification system for crop/weed discrimination without segmentation. In: IEEE Winter conference on Applications of Computer Vision (WACV); 2014. pp. 1142-1149

Chapter 3

Trends of Engineering Systems Evolution and Agricultural Technology

Ivan Mašín and Michal Petrů

Additional information is available at the end of the chapter

http://dx.doi.org/10.5772/intechopen.71481

Abstract

The new products are certainly decisive for achieving the business success of companies involved in the design and production of agricultural technology. Reducing the risk in the development and introduction of new technical products is the goal of analyzing the evolution of technical products. Effective innovation engineering procedures in the conceptual design phase do not use deductive methods such as brainstorming, but use more advanced methods with varying degrees of detail to describe the identified trends in the evolution of technology. In the case of this chapter, we will focus on so-called trends of engineering systems evolution. They describe natural transitions of the engineering system from one state to another, and are generally valid for all engineering disciplines. These are guides to the directions of development and their individual development phases, which should keep track of innovated products (through innovation, improvements and combinations of successful systems and technologies) so that users' needs are met more. Trends are generally the basis of modern technological forecasting and strategic planning. Unlike conventional forecasting methods, knowledge of trends can more accurately predict the problems associated with the introduction of new technologies and thus increases the probability of success of the chosen solution.

Keywords: evolution, trend, engineering, system, agriculture, technology

1. Introduction

Within the traditional approach to analyzing trends in the development of technical products, various procedures and methods [1–8] are used. We can divide these traditional techniques into five groups: techniques based on experience, assessment, and intuition; market research techniques; time series techniques; techniques using regression analysis; and other quantitative techniques. Information input for the majority of these methods is usually a subjective feeling and intuition, which of these methods makes a real mix of science and art. It is clear that the methods of technology forecasting should more include objective rules of the development of technical systems.

Because different technical systems overcome the "same types" of problems, their general development over time is "typical." These typical trends of engineering system evolution have been identified on the basis of a broad analysis of patent databases and historical trends in technology development [9–13]. Trends occupy a special place in the innovation science and engineering field, as they offer a view of the technical system from a variety of time perspectives—from the past through the present to the far future. They have a great potential for problem solving because they describe what happened in the past to successful technology and because they are leading the developer to what is likely to happen in the future for these technical systems. Therefore, prediction or, at least, a non-deductive view of what is likely to benefit the direction of development in a specific industry from the point of view of the system and its components can be used. In connection with the term "trend," it is also important to point out the terms "law" and "pattern." According to [13], the trend is describing general evolution directions of system components and can be graphically presented as a vector. "Laws of technical system evolution" only give general description of the links between phenomena. "Evolution pattern" is the specification of some trend (route map) [13]. Trends are of a statistical nature and cannot be understood as being filled in the last letter [14]. Of course there are frequent exceptions and deviations in which trends will not apply, for example, in a surprising technological discovery, fashion trends not based on logic or in situations where users' needs are not in line with technological developments.

Evolutionary trends represent, by analogy, external manifestations of "natural selection" in the world of technology, because even technical systems are struggling to use as well as biological species for survival sites. Systems that have survived their "birth" (idea, invention) and better met the demands of human society (i.e. high performance, low costs, and little impact on the environment) have won the evolutionary competition. As an example from agriculture, we can mention the hand scythe tool (**Figure 1**). In a traditional shape, scythe is perfectly adapted to carry forces from the human body and arms for effective grass mowing or grain mowing. However, the evolution of this agriculture tool has lasted more than 2000 years, and its development, for example, from the shape and material viewpoint still continues [15, 16].

Formally, trends or patterns of engineering system evolution are systemized guides and a description of the "winning" transitions from one development phase to the next that allowed the technical solutions to occupy and maintain a leading position in the market. Knowledge of trends and patterns can also be used in the so-called evolutionary analysis, which aims to obtain information on general directions for improvement of an innovative object or to formulate the correct tasks for the transition from one development phase to the other in the direction of the subline of evolution.

According to [10], trends of engineering system evolution are 11, and together they form a hierarchical system. From the two main trends (the S-curve development and the trend of increasing ideality), the other trends are derived (**Figure 2**).

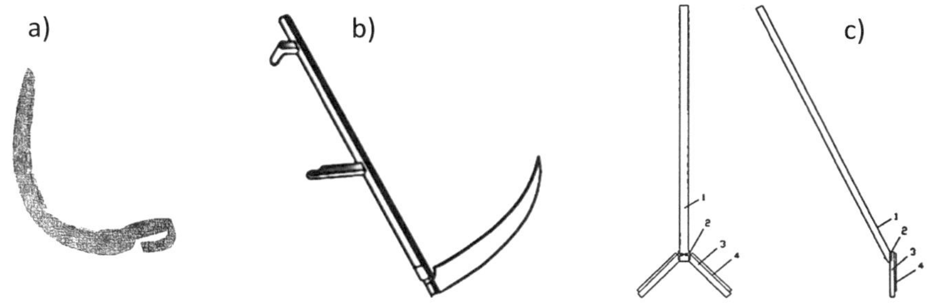

Figure 1. Hand scythe tool shape evolution ((a) ancient Egyptian sickle, (b) traditional European hand scythe, (c) newly patented Asian bidirectional scythe [16]).

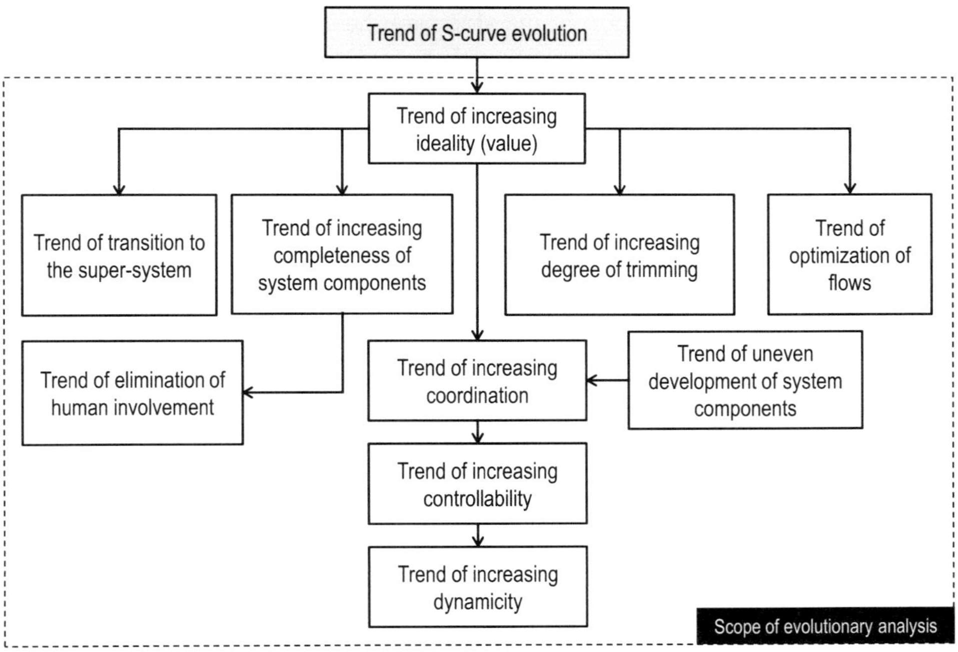

Figure 2. Hierarchical system of trends of engineering system evolution (adapted from [10]).

In the next section of this chapter, we will briefly focus on selected trends of engineering system evolution [10] from the agriculture technology viewpoint.

2. Following trends in the agriculture technology field (brief analysis)

Trends of engineering system evolution (TESE) can be a powerful tool for technology forecasting, innovation strategy development, and technical product innovation [17]. Related evolutionary analysis is an analytical-synthetic method, the aim of which is to look for the directions of innovation of the analyzed technical systems. This analysis usually takes place after identifying the shortcomings of current technical systems and technologies. In the analysis we are gradually considering the application of selected trends relevant to the analyzed field of technology. If, for example, there are no human-technology relations in the analyzed case, it is possible to exclude from the evolutionary analysis the trend of elimination of human involvement. The result of the evolutionary analysis is to assess the currently achieved stage of development of the given technical (sub)system and formulate recommendations and tasks for further innovation in the direction of successive phases or steps of a specific trend of engineering system evolution [10].

S-curve refers the shape of the logistic function used to illustrate the diffusion of innovations in the technology life cycle [18]. Since the trend of S-curve evolution mainly reflects the result of technological innovations expressed in the context of changes in the main parameters of technical systems, this trend is not considered in evolutionary analysis, but it is used at the very beginning of the innovation process at the stage of setting innovation targets and benchmarking studies (**Figure 2**).

Trend of increasing ideality (value) is the driving force for engineering system innovation along the S-curve, as well as driving force for diffusions of innovations. Altshuller [9] and [10, 13, 19] formulated ideality similarly to value as:

$$I\,(V)=\Sigma F/\Sigma C \tag{1}$$

where I (V) is engineering system ideality or value, ΣF is total functional capabilities or performance of useful function(s), and ΣC is total costs of performing function(s). System ideality can be increased by following principal approaches (see **Figure 3**):

1. $\Sigma F\uparrow;\ \Sigma C\approx \text{const} \Rightarrow I(V)\uparrow$
2. $\Sigma F\uparrow\uparrow;\ \Sigma C\uparrow \Rightarrow I(V)\uparrow$
3. $\Sigma F\uparrow;\ \Sigma C\downarrow \Rightarrow I(V)\uparrow$
4. $\Sigma F\approx \text{const};\ \Sigma C\downarrow \Rightarrow I(V)\uparrow$
5. $\Sigma F\downarrow;\ \Sigma C\downarrow\downarrow \Rightarrow I(V)\uparrow$

According to this trend, engineering systems increase their ideality, e.g. if their weight, dimensions, and energy consumption are decreasing (ideally close to zero) but their ability to perform functions is not diminished. In other words in the course of engineering system evolution, the ideality increases due to the increasing relation of system's functional potentialities and total cost of its creation and operation [10].

The limit ideal system is when cost is zero and/or functionality is infinite. Popularly speaking, "the technical system does not exist, but its main function are further provided." Functions in this case are the interactions of subject and objects, and they are described in the specific form as "subject – action (verb) – object." Agriculture is distinguished from other fields of

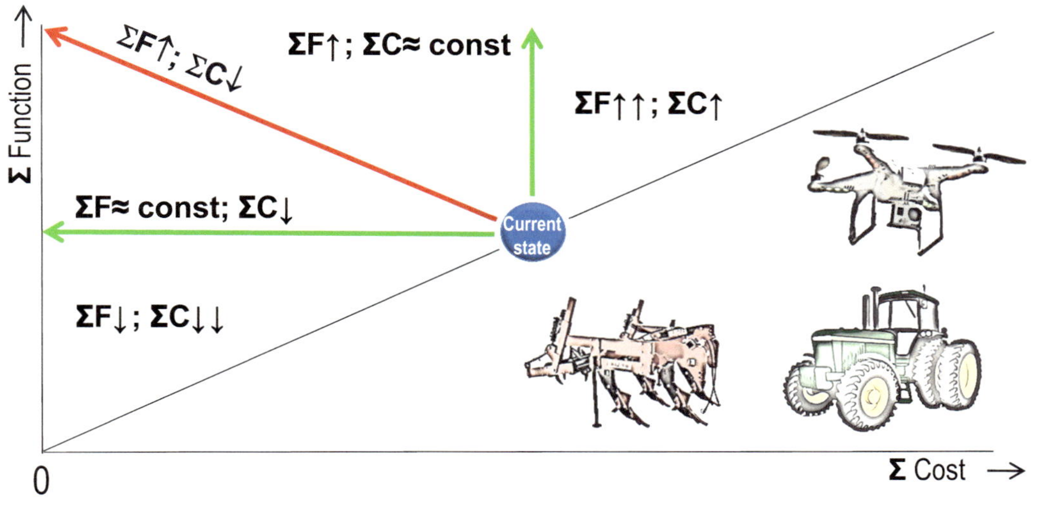

Figure 3. Possibilities to follow the trend of increasing ideality (value).

technology because it provides useful and harmful functions by system components as soil or animals. An example of this trend following toward an ideal technical system in the field of agriculture is shown in **Figure 4**.

The trend of increasing degree of trimming is that the number of components decreases during the development and innovation of the technical systems without compromising the functionality of the system (**Figure 5**). It is clear that one of the ways to reduce costs is to remove several components from the system. If we moreover succeed in preserving (or even increasing) the functional potential of the system, its idealness will naturally increase.

Maintaining functionality is ensured by following three trimming rules focused on redistributing useful functions of eliminated components to the remaining components or by transferring these useful functions to supersystem elements:

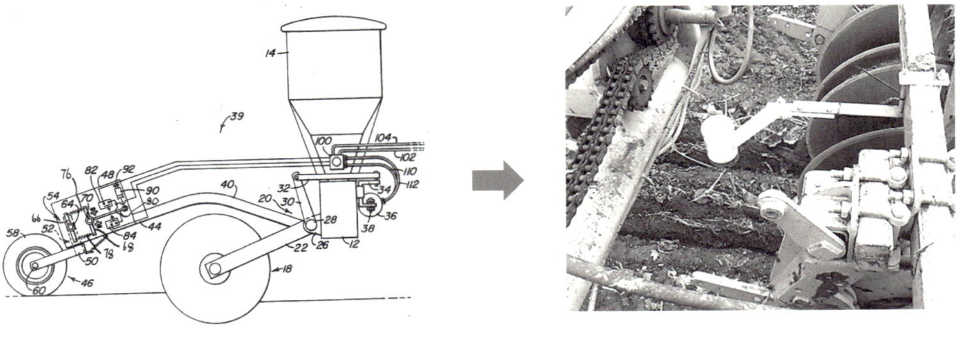

Figure 4. Mechanical depth control by wheels or rollers and)non-contact depth control by sonar sensors (controller) with lower weight and smaller dimensions [20, 21].

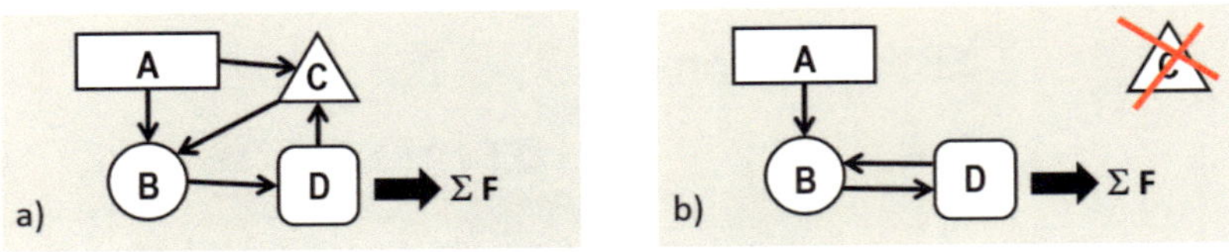

Figure 5. Trimming implies removal one or more components from original system (a – original system, b – improved system after trimming).

- trimming rule #1: function carrier (subject) can be trimmed if we remove the object of its useful function;
- trimming rule #2: function carrier (subject) can be trimmed if the object of function performs the useful function itself; and
- trimming rule #3: function carrier (subject) can be trimmed if another component performs its useful function.

In **Figure 6,** the trend of a higher degree of trimming can be illustrated by the example of hydroponic or aeroponic crop cultivation [22, 23]. By removing (trimming) the soil, many soilborne pests and diseases can be eliminated. This innovation also eliminates problems with nutrient availability and deficiencies caused by soil. As a result plants are healthier and grow much faster than plants in soil. Following the trend of a higher degree of trimming in a form of soil trimming enables to create and operate complex high-tech platforms for hydroponic or aeroponic crop growing in a closed and controlled environment. For example, a hydroponic growth system may be integrated into a programmable system providing for the growth of plants (**Figure** 7). An upper section of a system may include a lighting system able to vary lighting characteristics such as the intensity or spectral content of light provided to the plants and atmospheric systems to control the temperature, flow, or humidity of air around the plants that are mounted on an actuated platform. A control system may execute a program to control the available systems including the actuated above-plant platform to programmably control the height or heights of the systems above plants [24]. Such hydroponic or aeroponic high-tech systems also include advanced sensor systems. For example, the first sensor system measures one or more characteristics of a nutrient solution; a second sensor system measures one or more characteristics of an environment of a plant and a network device including a communication interface to the first sensor system and a communication interface to the second sensor system. The network device may be configured to transmit measurements from the sensor systems through a wireless network to a remote device or database in the Agriculture 4.0 direction [25].

One of the strongest and most frequently applied trends is the **trend of transition to the supersystem.** This trend reflects the fact that in the course of its evolution, the systems exhaust their internal resources and subsequently the original system integrates with other (even alternative or competing) systems and continues its development in the so-called supersystem (freely understood the supersystem as a "surround" of the system). As a rule, the functions

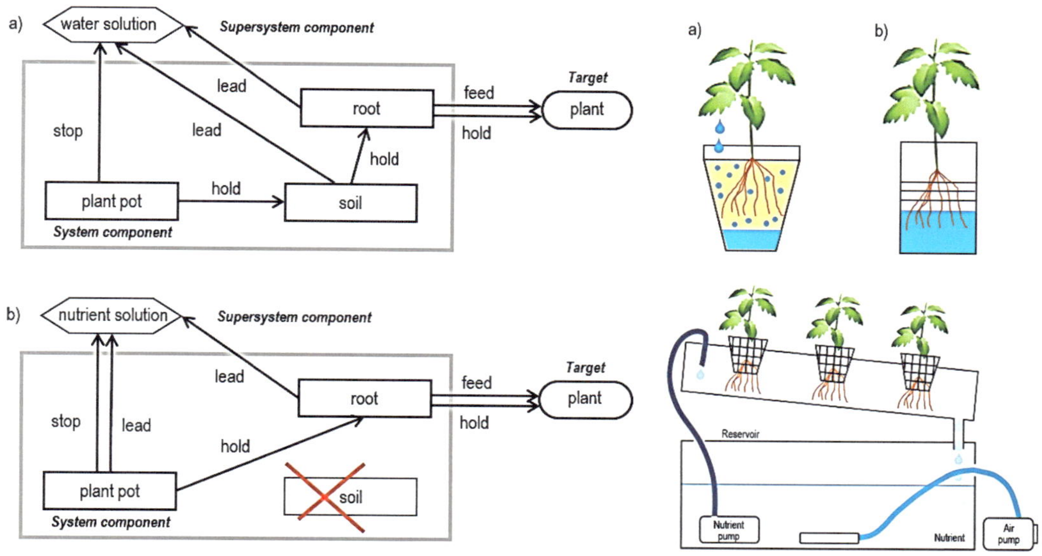

Figure 6. Functional model of traditional plant growing (a) and trimming model for hydroponic plant growing (b).

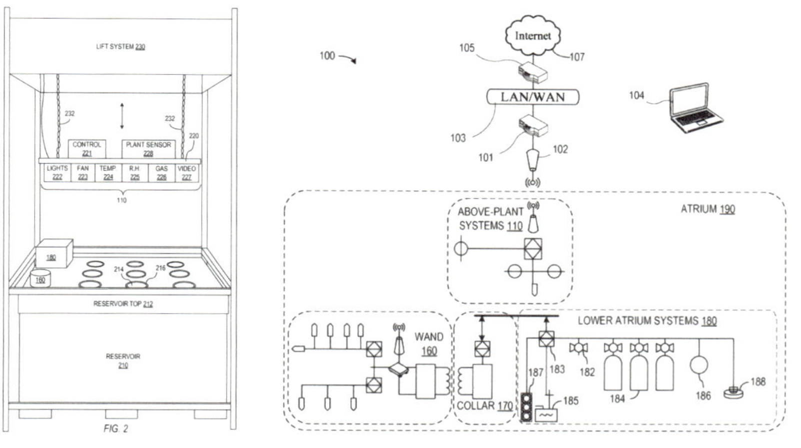

Figure 7. Hydroponic system with actuated above-plant platform [24].

and quality of the old system are passed on to the new one, i.e., the new system does not originate without the old system. The trend has four subtrends [10]:

- subtrend #1: the parameters of the integrating system are increasingly different from the original system parameters.
- subtrend #2: the main functions of integrating system are increasingly different from the original system main functions.

- subtrend #3: the level of integration between the original system and the integrating system is deepening.
- subtrend #4: the number of systems that integrate with the original system is increasing.

Example of the subtrend #1 in the agriculture field is illustrated in **Figure 8** by multipurpose agricultural unmanned helicopter as an agent of intelligent agriculture technology. The transition from one development phase (vehicle) to the next stage (rotorcraft) is triggered by the rapid development of modern agriculture, where traditional spraying or transporting methods no longer meet current requirements of precision agriculture.

The **trend of increasing efficiency of substance, energy, and information flows** is that the evolution of technical systems that contain substances, energy, and information flows is progressing toward increasing the efficiency of the use of these flows [10]. Under normal circumstances, any flow transformation (transformation of a substance from one state into another, changes in energy types, changes in the content of information, etc.) is accompanied by losses, deceleration, and delay. In fulfilling this evolutionary trend, the stream with several transformations of energy turns into a homogeneous flow. Ideally, there should be no transformation at all, and all components of the flows should be at once in the form that is needed for their utilization [10]. This trend is illustrated by example of the straddle electric tractor (**Figure 9**) showing transition of the agricultural ground transport from combustion engines with several energy transformations to electric propulsion and battery pack that will fully power even large tractors for agriculture.

The **trend of increasing coordination** is reflected in the evolution of technical systems by gradually coordinating the "behavior" of the system components and consequently coordinating the "behavior" of the supersystem [10]. Coordination is also understood as the choice of one parameter with respect to the value of another parameter. This "driving" parameter value can be preselected (e.g., when manufacturing the agricultural machine) or in the process of its operation (e.g., during harvesting or weed-destroying processes). For example, when developing the outer shape of a system, the shape of the developed system must be coordinated with the shape, properties, and movement of the objects that interact with the system (an example of this trend being the standardization of the dimensions of the interconnected parts or the ergonomic agricultural tool handle solution). From this trend viewpoint, we can observe different trend mechanisms—coordination in shape, coordination in rhythm, coordination of materials, coordination of action (**Figure 10**), coordination of parameter, self-coordination, etc., [10].

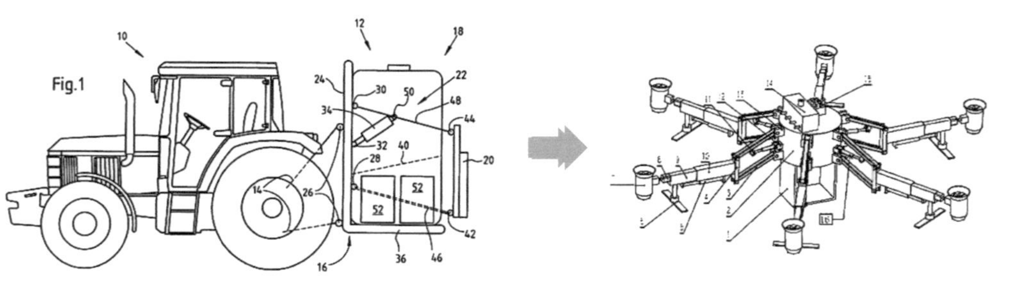

Figure 8. The transition to an integrated system with better parameters than the parameters of the original engineering system [26, 27].

http://dx.doi.org/10.5772/intechopen.71481

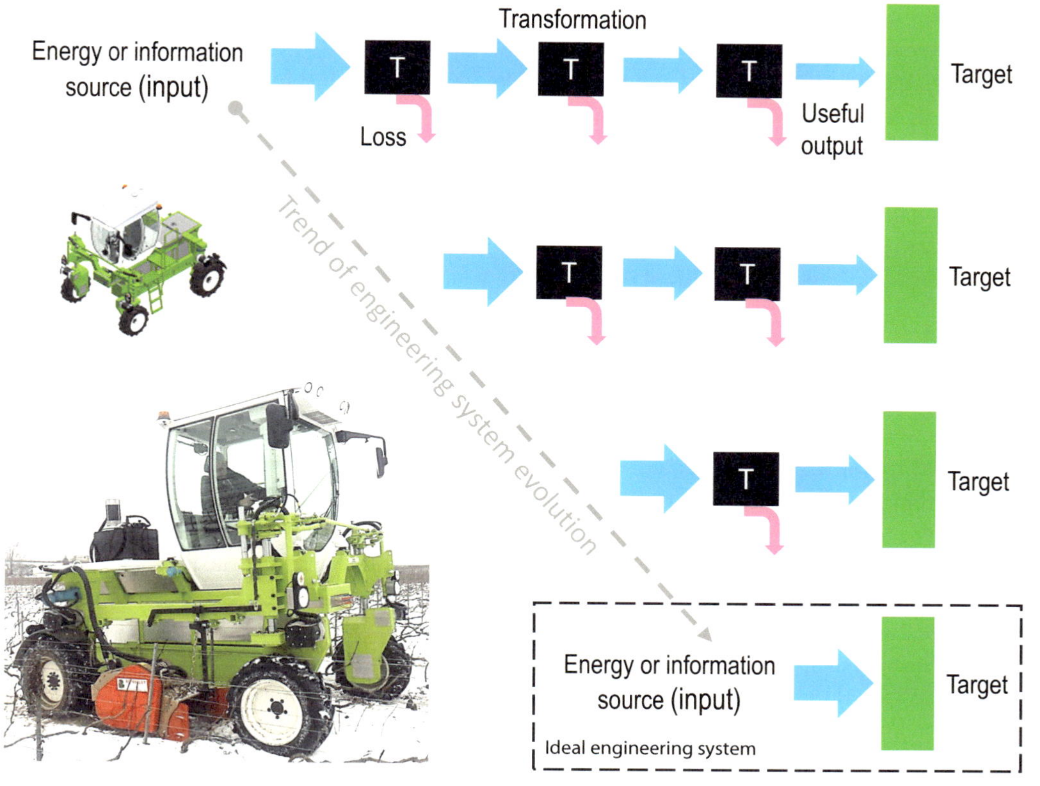

Figure 9. Enhancement of conductivity of useful flows through reduction in the number of flow transformations (one of the possible fulfillments of the trend in the form of electric straddle tractor, T4E [28]).

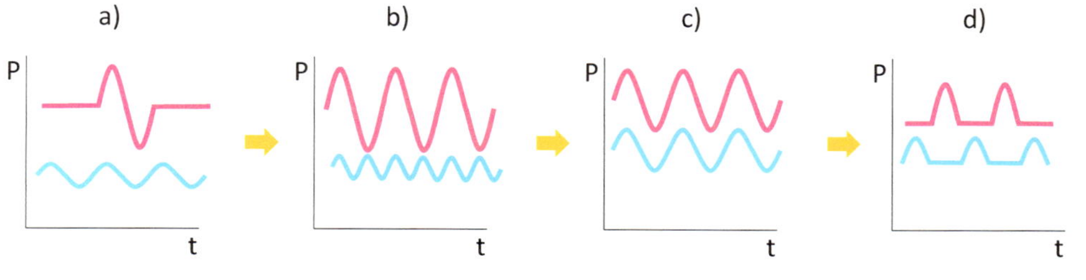

Figure 10. Trend of increasing coordination of engineering systems ((a) incoordinated actions, (b) partly coordinated actions, (c) coordinated actions, (d) interval actions).

The trend of increasing coordination in the form of coordinated actions can be illustrated by example of a system for speed-based coordinated control of agricultural vehicles when on-the-go unloading is utilized (i.e., the harvester is in motion) (**Figure 11a**). This control system [29] includes a controller communicatively coupled to the first transceiver, wherein the controller automatically controls the speed of the haul agricultural vehicle by determining a desired position and a desired speed of the haul vehicle based at least in part on the first determined position and the first determined velocity of the target vehicle, instructing an automated speed control system to establish the ground speed of the haul vehicle to reach the target position and instructing the automated speed control system to control the ground speed of the haul vehicle to maintain the target position, including during turning of the target and haul vehicles. The transceivers may broadcast and receive radio waves within a frequency range of about 1 GHz to about 10 GHz. In addition, the first and second transceivers may utilize any suitable communication protocol, such as a standard protocol (e.g., Wi-Fi, Bluetooth, etc.) or a proprietary protocol [29]. Another example of this trend may be wireless networking of agricultural machines in a collaborative agricultural process (**Figure 11b**) [30].

The **trend of increasing dynamicity** lies in the fact that during the development of the engineering system (components), the flexibility, dynamism, or adaptability increases. The development of engineering systems in the direction of this trend proceeds from an initial rigid structure with unchanging parameters to a more flexible and variable structure with more degrees of freedom, with adjustable parameters and a working mode adapted to changes in the external environment. For example, engineering system that has rigid components is poorly adapted to operating conditions. For this reason, the evolution of rigid components becomes more flexible and dynamic. First, a single joint, and then multiple joints, are inserted into the solid monolith. Moving from joints to flexible fabrics, the system gains flexibility and adaptability. Fluids or gases are even more adaptable. Finally, the system is converted to some of the physical fields that have the best controllability (**Figure 12**) [10].

The trend of increasing dynamicity in the agriculture technology field can be illustrated by innovations realized within precision agriculture (PA) that is based on observing, measuring, and responding to crops or soil variability. Proactive decisions are made with the support of

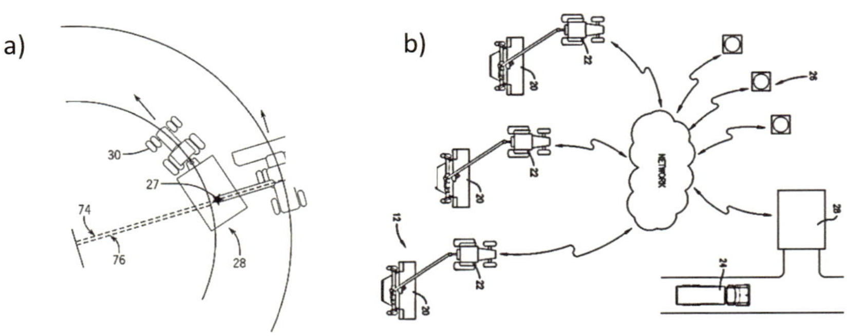

Figure 11. Automatic coordination of the agricultural vehicles speed (a) and synchronization of agricultural vehicles with the help of wireless networking (b) [29, 30].

Figure 12. Trend of increasing dynamicity of engineering system component.

Agriculture 4.0 tools and techniques as sensor systems, robotics, computational techniques, positioning systems, or big data to manage variability and control site-specific applications. Progress in this trend direction can be documented on the development of harvesting robot grippers whose design deviated from hard-to-handle robot hands with rigid fingers to design with flexible fingers or cushioned finger design (**Figure 13**).

The **trend of elimination of human involvement** in engineering system trend lies in the fact that during evolution, the number of functions performed in the system by human is reduced [10]. This trend is particularly relevant for systems where standard subsystems such as the workpiece, transmission, power source, and control system are omitted and initially do not exist (**Figure 14**). This is a special case of the **trend of increasing completeness** of engineering systems, because a man is often an element on which it is usually easiest to transfer the functions, which yet cannot be performed by the system.

Fulfilling this trend in agricultural technology can be documented on the concept of an autonomous cableless tractor [35] or on the unmanaged robotic platform for performing multiple functions in agricultural systems [36] (**Figure 15**).

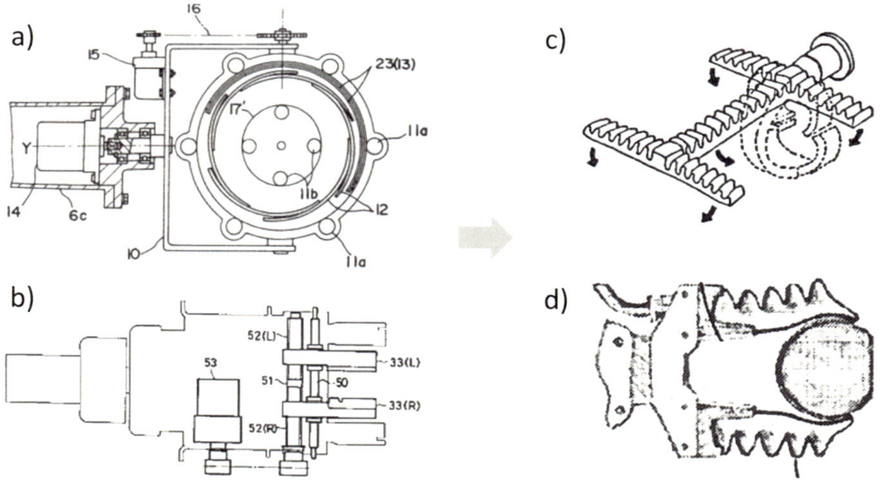

Figure 13. Dynamicity increasing of harvesting robot grippers ((a) [31], (b) [32], (c) [33], and (d) [34]).

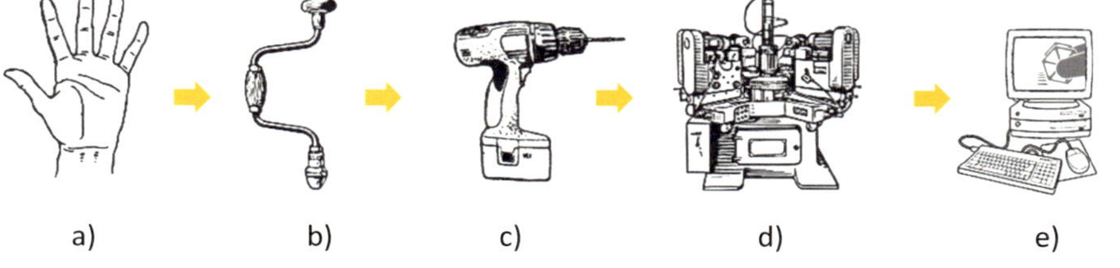

Figure 14. Trend of elimination of human involvement in engineering systems (human roles: (a) man alone, (b) tool, (c) energy and drive, (d) control and supervision, (e) only decision-making).

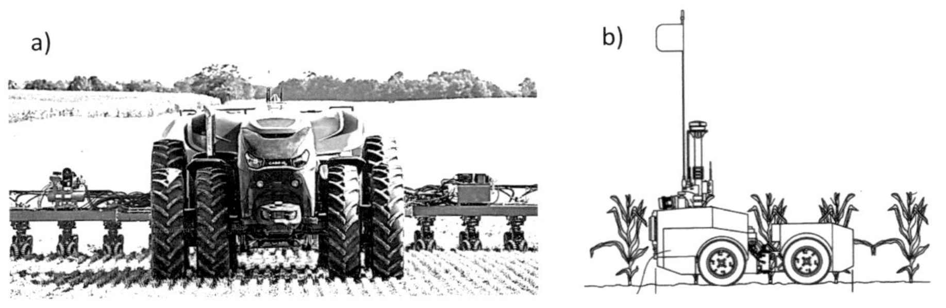

Figure 15. Trend of elimination of human involvement in engineering systems: concept of autonomous tractor (a) and robotic platform (b) [35, 36].

3. Technology forecasting using systematic creativity methods

The paradigm of innovative companies is the transition to new, market-successful, types of products. However, market success must be thoughtfully directed, i.e., it is necessary to read, respectively, to forecast the future. All traditional forecasting techniques (market research techniques, Delphi method, time series analysis, regression analysis, etc.) are aimed at finding key issues for shareholders, investors, and business owners before launching development work and putting new technical products on the market "What is the business potential of the new product?" Input data are typically industry and market information, experience, subjective feelings, and intuitions, which make these methods a mix of science, randomness, and art. In order to solve the prediction task more objectively, it is possible to use two creative approaches extending traditional technological forecasting

- TRIZ technological forecasting—what changes should be made to move our product or process to a further developmental position in the specific trend of the technical systems?
- Directed evolution (**Figure 16**)—which development scenario should be selected from the identified set of scenarios to make the product, service, or process successful?

Since directed evolution (DE) can be considered more complex, let's take a brief look at its principles. The coauthors of directed evolution method B. Zlotin and A. Zusman [8] continued in the classical TRIZ forecasting (which alone is not enough to manage the development and evolution of products) and added evolutionary sources to Altsuller's innovation principles. Directed evolution is then based on the following five theoretical backgrounds (postulates) for creating scenarios of future development:

1. Application of system development models (lines). Most manmade technical systems follow rather predefined models than they would be a result of random phenomena. This means that a strategy based on the proactive approach of the eight identified development models and the resulting development lines can maximize profits and reduce the cost of technical or other innovation.

2. Market-oriented direction of development. Most manmade technical systems follow tor customer satisfaction. It means that they evolve in the direction of increasing ideality.

3. Consumption of resources for system development. The development of the system is based on the consumption of internal resources, sources from surrounding systems and the surrounding area. In the initial stages of the S-curve, easily available sources are utilized. These sources are later replaced by complex or hidden sources. New types of products will usually appear when new sources are discovered (structure or material).

4. Priority of long-term forecasts. Prior to improving system parameters and resources, long-term general development of the technology is preferred, including the inclusion of new generations and fundamental discoveries.

5. Application of system development alternatives. The engineering system always has more than one possible path to the next stage of development (according to various sources). The most profitable new system is usually the one that is launched first, which attracts most of the financial and human resources [8].

The basis of the DE is the diagnosis that follows the initial formulation of the problem and the collection of adequate data. The aim of this diagnosis is to identify the possible directions of evolution (development scenarios) of the given engineering system and to formulate all the problems and tasks that must be accomplished in order to fulfill this direction of evolution (development). Performing DE diagnosis means, for example:

- Compare historical data with trends and development lines.
- Identify missing and future development steps that are understood as innovative opportunities.
- Identify bad development trends.
- Identify unsolved contradictions.
- Extrapolate development directions (lines) for a given technical system [8].

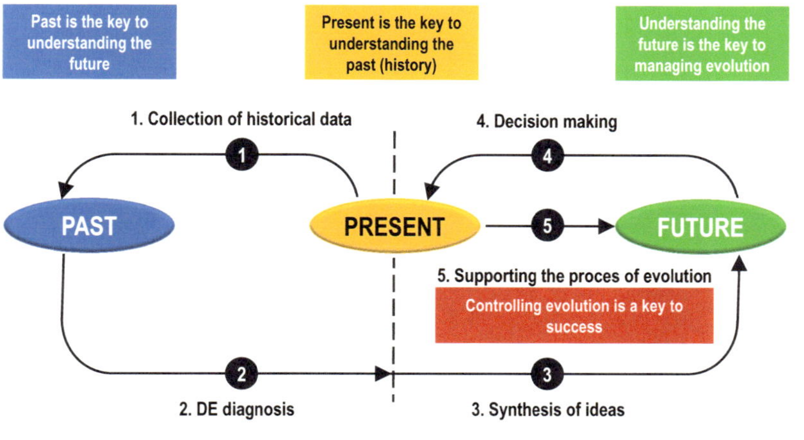

Figure 16. General schema of directed evolution [8].

The result of the DE is the concepts of innovative products that correspond to both identified scenarios (trends) of the development of a given technical (or social [8]) system and concepts based on the resolution of exposed contradictions. Prognoses created using systematic creativity methods can provide management and engineers with important and more reliable input into the strategic planning process because they are based on objective patterns in the development of technical systems. These predictions can be used not only to better understand future threats and present opportunities but also to actively influence the future as a contrast to passive and reactive problem solving, because the best way to predict the future is to invent it.

4. Conclusion

Trends of engineering system evolution (TESE) can be a powerful tool for agricultural technology innovation and forecasting. The goal of TESE is to provide objective and analytical tools for problem identification of engineering systems, assist in forecasting the future evolution of agricultural technology, and recommend solutions for innovation of engineering systems based on their evolutionary stage. Applying evolutionary analysis to the engineering systems guarantees a supply of novel ideas, trimming recommendations, and conceptual directions.

Acknowledgements

The results of this project LO1201 were obtained with co-funding from the Ministry of Education, Youth and Sports as part of targeted support from the "Národní program udržitelnosti I" program.

Author details

Ivan Mašín* and Michal Petrů

*Address all correspondence to: ivan.masin@tul.cz

Institute for Nanomaterials, Advanced Technologies and Innovation, Technical University of Liberec, Liberec, Czech Republic

References

[1] Kahn KB. New Product Forecasting. An Applied Approach. New York: M.E. Sharp; 2006

[2] Christensen CM. Exploring the limits of the technology S-curve. Part 1: Component technologies. Production and Operations Management. 1992;**1**(4):334-357

[3] Courtney H, Kirkland J, Viguerie P. Strategy under uncertainty. Harvard Business Review. 1997;**75**(6):68-79

[4] Garcia ML, Bray OH. Fundamentals of Technology Roadmapping. Sandia National Laboratories: Albuquerque; 1997

[5] Phaal R, Farrukh C, Probert D. Roadmapping for Strategy and Innovation: Aligning Technology and Markets in a Dynamic World. Cambridge: Institute of Manufacturing - University of Cambridge; 2010

[6] Miles I, Saritas O, Sokolov A. Foresight for Science, Technology and Innovation. Cham: Springer International Publishing Switzerland; 2016

[7] Daim TU, Pizzaro M, Talla R. Planning and Roadmapping Technological Innovations. Cham: Springer International Publishing Switzerland; 2014

[8] Zlotin B, Zusman A, Zainiev G. An Application of Directed Evolution. Southfield: Ideation International Inc.; 2012

[9] Altshuller GS. Creativity as an Exact Science. New York: Gordon and Breach Publishing; 1984

[10] Lyubomirskiy A, Litvin S. Trends of Engineering Systems Evolution - Guide. Boston: Gen 3 Partners; 2003

[11] Gadd K. TRIZ for Engineers: Enabling Inventive Problem Solving. Chichester: Wiley; 2011

[12] Fay V, Rivin E. New Product Development Using TRIZ. Cambridge: Cambridge University Press; 2005

[13] Shpakovsky N. Tree of Technology Evolution. Ways to New Business Opportunities. CreateSpace: Moscow; 2016

[14] Rubin M. Myths about the Trends of Engineering Systems Evolution [Internet]. 2009. Available from: http://www.temm.ru/en/section.php?docId=4386 [Accessed: 20.8.2017]

[15] French M. Invention and Evolution. Design in Nature and Engineering. 2nd ed. Cambridge: Cambridge University Press; 1994

[16] Zhou H. Patent CN202425318. Detachable Bidirectional Scythe; 2012

[17] GEN TRIZ. Methodology. Technology trends [Internet]. 2016. Available from: https://www.gen-triz.com/technology-trends/ [Accessed: 21.8.2017]

[18] Roger RM. Difussions of Innovations. New York: Free Press; 2003

[19] Lyubomirskiy A. Ideality equation. TRIZ Journal [Internet]. Jul 2015;**27**. Available from: https://triz-journal.com/ideality-equation/ [Accessed: 21.8.2017]

[20] Hamm N, Rosbak J. Patent EP0057571A1. Agricultural Implement with Automatic Depth Control, 1982

[21] RiteDepth – Automatic Depth & Distance Controller [Internet]. Available from: http://greentronics.com/products/automatic-depth-controller/ [Accessed: 22.8.2017]

[22] Umpstead N. Patent US2016135395A1. Modular Hydroponic Growing System, 2016

[23] Orff D. Patent US2016227719A1. Aeroponic System, 2016

[24] Wong S, et al. Patent US2016128289A1. Hydroponic System with Actuated Above-Plant-Platform, 2016

[25] Chen T. Patent US2016255781A1. Wireless Sensor Systems for Hydroponics, 2016

[26] Arnaud H. Patent DE10148422. Pesticide Sprayer has Tank Mounted on Trailer Behind Tractor, Spray Bar Being Mounted on Control Bars Raised and Lowered by Piston and Cylinder Unit and which pass through Aperture in Tank, 2003

[27] Xueheng Z. Patent CN106114832. Multipurpose Agricultural Unmanned Helicopter, 2016

[28] Kremer Energie. T4E - Straddle Electric Tractor [Internet]. Available from: http://www.kremer-energie.fr/en/pages/technical-specificities-of-t4e.html [Accessed: 09-02-2017]

[29] Ray BR, et al. Patent US2017192419A1. System and Method for Speed-Based Coordinated Control of Agricultural Vehicles, 2017

[30] Schmidt KK, et al. Patent US2017083026A1. Isobus Wireless Networking of Agricultural Machines in a Collaborative Agricultural Process, 2017

[31] Takashi T. Patent US4663925A. Fruit Harvesting Robot Hand, 1987

[32] Nagai H, Arima S. Patent JPH05168333. Fruit Harvesting Robot, 1993

[33] Galloway K, et al. Patent WO2017127497A1. Soft Robotic Actuators and Grippers, 2017

[34] Lessing J, et al. Patent US2017203443A1. Soft Robotic Grippers for Cluttered Grasping Environments, High Acceleration Movements, Food Manipulation, and Automated Storage and Retrieval Systems, 2017

[35] CNH Industrial. CNH Industrial Brands Reveal Concept Autonomous Tractor Development [Internet]. Available from: http://www.cnhindustrial.com/en-US/media/press_releases/2016/august/pages/NH_Industrial_brands_reveal_concept_ autonomous_ tractor_development_Announcement.aspx [Accessed: 2.9.2017]

[36] Kavender-Bares K. Patent US20123424A1. Robotic Platform and Method for Performing Multiple Functions in Agricultural Systems, 2017

Section 2

Industrial Automation Examples in Agriculture

Chapter 4

Review of Variable-Rate Sprayer Applications Based on Real-Time Sensor Technologies

Zhihong Zhang, Xiaoyang Wang, Qinghui Lai and
Zhaoguo Zhang

Additional information is available at the end of the chapter

http://dx.doi.org/10.5772/intechopen.73622

Abstract

Precision variable rate spray is one of the research hotspots in the field of modern agriculture spraying applications. Variable rate spraying of the canopy allows growers to apply adjusted volume rate of pesticides to the target, based on canopy size, and to apply plant protection products in an economical and environmentally sound manner. In the field of pesticide application, knowledge of the geometrical characteristics of plantations will guarantee a better adjustment of the dosage of the agrochemicals applied. This technology is integrated with intelligent real-time sensors, which have a high potential for agricultural precision spray applications. This book chapter presents the foundations and applications in agriculture of the primary systems used for real-time spray target detection of the geometrical characterization of tree plantations. Systems based on infrared, ultrasonic, light detection and ranging (LIDAR), and stereo vision sensors were discussed, respectively, on their performances to detect spray targets. Among them, laser scanners and stereo vision systems are probably the most promising and complementary techniques for achieving three-dimensional (3D) pictures and maps of plants and canopies. The advantages of data fusion applied in real-time target detection and its accuracy in density estimation of the plants were stressed.

Keywords: variable-rate sprayer, infrared sensor, ultrasonic sensor, stereo vision sensor, LIDAR sensor, data fusion

1. Introduction

In agriculture, chemicals are often essential for crop protection. Pesticide spray applications have facilitated high-quality and abundant products for ornamental nurseries and orchards.

However, despite these achievements, conventional sprayers are grossly inefficient because the same amounts of chemicals are discharged continuously in the field regardless of the plants present, canopy structure, or leaf foliage density. Canopies are spatially variable, and a uniform dose may not be adequate for the entire orchard. Since plants are often either over or under sprayed, resulting in environmental pollution issues and inadequate pest control. Besides, growing pressure from farmers, environmental organizations, and public opinion are encouraging lawmakers to try to reduce pesticide losses to the environment. Spraying at an adequate volume application rate on a site-specific basis would help reduce the amount of agrochemicals used in the framework of precision horticulture and precision fruticulture.

Canopies are spatially variable, and knowing the structural characteristics of the canopy is a crucial consideration for improving the efficiency of the spray application process for tree crops. The introduction of electronic systems in the development of new equipment helps to reduce both operating and environmental costs by optimizing the efficiency of the pesticide treatments. For instance, machines that spray only in the presence of plants, not in the gaps between them, have already been developed for cabbage vegetable crops [1], peach, and apple tree cultures [2]. An essential goal for orchard and vineyard spraying systems is a real-time adjustment of the operating parameters according to the target density, with the aim of keeping the droplets in the canopy, thus improving spray deposition and reducing spray drift.

Therefore, to reduce pollution during spray operations, interest in variable-rate spray technology is growing. A promising solution is the new intelligent variable-rate spray technology that automatically controls spray outputs to match plant presence, canopy characteristics, and travel speeds. This currently available technology can reduce pesticide use and off-target losses, and thus its use will benefit farmers, consumers, and the environment. Advances in sensing and detection technologies may facilitate precision autonomous operations that could improve crop yield and quality while saving energy, reducing workforce, and being environmentally friendly. Real-time sensor and control systems on sprayers are necessary to achieve a uniform spray deposit on the crop canopies and to reduce spray losses. These sensor systems are based on different kinds of physical principles, which may allow efficient monitoring of the canopies. The premise of precision spraying is the detection of the characteristic information of the target plant, which is the foundation and basis for the spraying. However, obtaining accurate data in an easy, practical, and efficient way is a significant problem to be solved. This book chapter will review the real-time sensor based on the precision variable spray method.

2. Infrared sensor-based detection technology

All objects with a temperature above absolute zero emit heat energy in the form of radiation. Infrared sensor is an electronic sensor that measures infrared light radiating from objects in its field of view. This technique works entirely by detecting infrared radiation emitted by or reflected from objects. An infrared detector utilization is in the automatic target detection system. Infrared sensor-detecting techniques have been adopted in automatic target-detecting orchard sprayers to discern targets and control the spraying system automatically. These sprayers can be commercialized easily due to the low price of infrared sensor detectors. Developed countries such as the USA, EU, and Russia are developing automatic target-detecting sprayers that utilize infrared imaging techniques [3–5]. Due to the problems related to infrared image processing, these sprayers remain in the experimental stage.

He et al. [6] designed a precision orchard sprayer based on automatic infrared target-detecting and electrostatic spraying techniques (**Figure 1**). The sensors are aimed at the top, middle, and bottom segments of the tree canopy to detect different shapes of fruit trees and provide signals to the control system. Experimental results show that the new automatic target-detecting orchard sprayer with an infrared sensor can save more than 50–75% of pesticides, improve the utilization rate (over 55%), control efficiency, and significantly reduce environmental pollution caused by the pesticide application.

Bargen et al. [7] designed a red/near-infrared reflectance sensor system for detecting plants. These reflectance characteristics have been determined using spectra-radiometry technology. Detection of plants is possible based upon the distinct reflectance characteristics of plants, soil, and residues. Optical filters were used to select the spectral bandwidth sensitivities for the red and near-infrared ray photodetectors. The reflectance values were digitized for incorporation into a normalized difference index in order to provide a stronger indication that a live plant is present within the field of view of the sensor. This sensor system was combined with a microcontroller for activating a solenoid-controlled spray nozzle on a single-unit prototype spot agricultural sprayer. Jiao et al. [8] designed infrared photoelectric switch and applied it to spraying on aspen. The experiment proved that infrared photoelectric switch attained the request of the design and reduced the cost of spraying. The interval of target identification was less than 0.3 m, and range of target identification was between 0.2 and 15 m. Adjustable work minimum pass spacing was less than 3.0 m. Jianjun et al. [9] developed an infrared detecting system consisting of integrated circuit for orchard automatic target sprayer. This system satisfied in detail the design requirements of stability, sensibility, compact volume, and anti-interference from environmental ray, and the detectable distance between the detector and the targets was variable from 0 to 6.15 m, and the space between two spraying targets was no more than 0.3 m.

Infrared detection technology in plant targeting is more applicable for dense and large target reflectors under high light intensity. It will get the best detector sensitivity near midpoint of detection distance and give better detection results for plants with high leaf reflectivity.

Figure 1. Photo of the automatic target-detecting orchard sprayer working in orchard [6].

However, when utilizing infrared target detection for plant pesticide spraying, the operation of an infrared detecting system for automatic target orchard sprayer was hard to work well in rough environment interference resulting from designed defects, including short detectable distance, complicated circuit, and the high cost of the automatic target detector. Although temperature and humidity have little impact on the detection results, plant appearance, light intensity, walking speed, and plant space have evident influences on detecting effect, especially the plants' appearance and light intensity. Plant density and light intensity are proportional both to detection distance and width. The speed of detector has a linear correlation with the minimum distance of individual plant efficiently distinguished. Plant space is monotonously correlative to detecting sensitivity [10]. Besides, due to the limitation of this sensor, the detection method based on infrared technology cannot detect the characteristic information such as the specific size and size of the target, that is, the qualitative calculation and analysis cannot be realized. Also, the detection process is easily exposed to external light influence [11, 12], and with the continuous growth of modern agricultural spray operation requirements, the technology has been gradually unable to meet the development needs.

3. Ultrasonic sensor-based detection technology

Another type of system is based on the use of ultrasonic sensors to measure distances quickly and automatically. These sensors have three essential elements: an emitter of ultrasonic waves, a chronometer, and a wave receiver. Their operation is based on determining the flight time of an ultrasonic wave from the point of emission to the point of detection after bouncing off an object. The potential application of ultrasonic sensor includes orchard management based on rapid quantification of tree volume. The information could be used in variable-rate application of agrochemicals within a grove. There was without spraying when there was no vegetation, half spraying when there was little vegetation in front of the sensors and full spraying when sensors detected the width of the canopy above a given threshold. This achievement led the way to a continuous variation of flow rate according to the variability of the canopy along citrus groves, vineyard, and fruit orchard rows [11, 13–15].

Different researches have been conducted for automatic measurement of canopy dimensions in groves. For decades, ultrasonic sensors have been employed in agriculture for different purposes [16, 17]. One of these applications is detection and ranging to obtain structural data from trees. The first advances in this field were related to the application of plant protection materials such as pesticides in different orchards. When dose adjustment according to canopy structure was proposed [18], some researchers began to design electronic systems for measuring canopy structural parameters. The first proposed systems to determine canopy volume used many ultrasonic sensors on a vertical mast [19] or mounted on the sprayer [20]. Because of the state-of-the-art of the application technologies, using this information in real time was not possible. The use of ultrasonic sensors has been reported only for the detection of canopy presence by [2, 21]. In this method, spraying was done exclusively when the canopy was in front of the sprayer. Another application was citrus trees spraying from constant given distance [18]. The nozzles were located on a movable arm, which follows the boundary of the tree according to data collected from sensors. Ultrasonic sensors were placed 50 and 75 cm apart. The same authors improved another sprayer that was able to spray with three different dosages according to width estimation of the canopy made by ultrasonic sensors [5].

In the USA, the performance of a sprayer prototype using ultrasonic sensors was tested by Giles et al. [2]. The system adjusted the flow rate of the sprayer to the canopy size variations measured by the sensors. The spray boom was divided into three sections each side, and these sections were independently turned on and off according to the readings of ultrasonic sensors, placed at different heights. Spray savings were reported, but there was also less spray deposition on some foliage areas when the control system was used. In the late 1980s, sprayer models appeared on the market, which were able to turn off the spray when there was a gap between trees [20]. It is beneficial for saving spray in young orchards or when there are wide gaps between trees, reducing the spray drift and the chemical cost. However, these systems do not account for variations in canopy shape, which are found in most of the orchards. More recently, another approach was made by Balsari and Tamagnone [22] with an ultrasonic control system mounted on a ducted air-assisted sprayer. In this case, the number of working nozzles could be adjusted to tree height, according to the readings of sensors placed at different heights. Tumbo et al. [23] proposed the use of ultrasound sensors to estimate the volume of citrus trees using the principle of time of flight to determine the distance to the target. Adopting the same system, Zaman and Salyani [24] proved that forward speed is not as important as tree density on volume estimation. Planas et al. [17] reported interferences between adjacent sensors spaced less than 60 cm apart. This method assumes the constant distance from the sensor to the tree center, and a small variation on this distance results in a large error on the final volume estimation. Balsari et al. [25] went one step further analyzing the crop identification system and concluded that there is a relationship between canopy density and its ultrasonic echo signal. Palleja and Landers [26] reported a low-cost system using four ultrasonic sensors and a microcontroller board to estimate the canopy density as a function of the ultrasonic echoes. It was tested as the growing season progressed and the data obtained highly correlated with the season, but they were not compared to actual canopy density.

Moltó et al. [5] developed a prototype to turn off the spray in the gap between two tree canopies and with the possibility of making up for the variation of canopy volume at the beginning and end of each tree (**Figure 2**), using the action of two electro valves at each boom section. An automatic sprayer has been developed that, using an electronic control system, adapts the dose of the product to the actual amount of leaf mass. This system is based on a cheap, 8-bit, conventional microcontroller that receives information about the tree shape from two ultrasound sensors and actuates through several electro-hydraulic valves mounted on a specially designed hydraulic circuit. The system allows spraying higher doses in the central part of the tree, where there is more vegetation in globular shaped canopies. Under the conditions of field test experiments, the system achieved savings of up to 37% of the product while maintaining the quality of the treatment. These savings depend on the size, shape, and distance between trees in each particular orchard.

Gil et al. [15] pointed out that target detection with ultrasonic sensors can be used to adapt the applied dose following the principles of the variable-rate technology. A multinozzle air-blast sprayer (**Figure 3**) was fitted with three ultrasonic sensors and three electro-valves, to modify the flow rate from the nozzles in real time, in relation to the variability of crop width. A constant application rate of 300 l/ha^{-1} was compared with a variable-rate application using the tree row volume principle at a 0.095 l/m^{-3} canopy. The total flow rate sprayed by the nozzles was modified according to the variations of crop width measured by the ultrasonic sensors. On average, 58% less liquid was applied compared to the constant rate application, with similar deposition on leaves with both treatments.

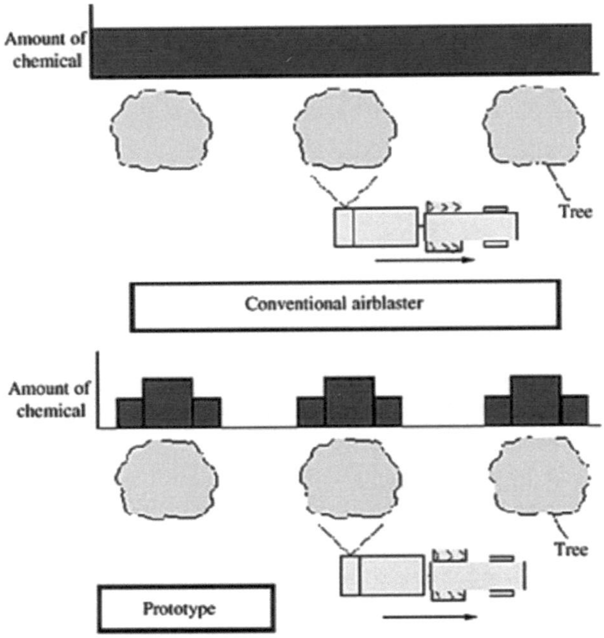

Figure 2. Chemical applied by a conventional sprayer and by the prototype [5].

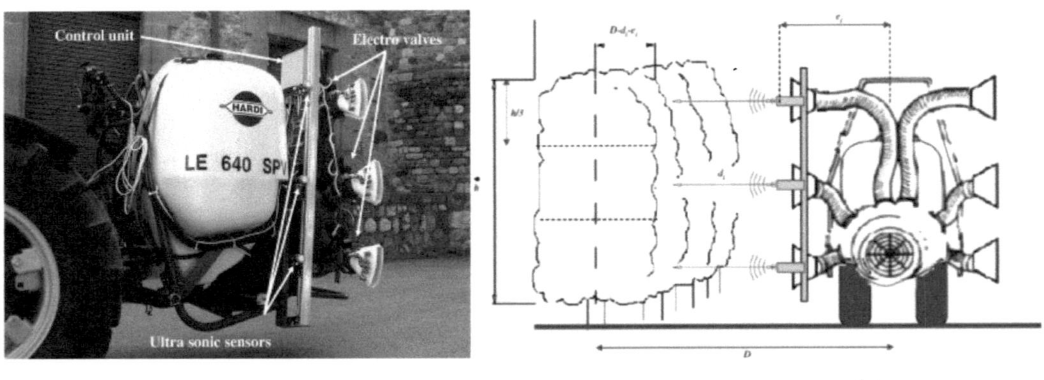

(a) (b)

Figure 3. (a) Sprayer prototype with ultrasonic sensors and electro-valves, (b) principle of operation of the prototype [15].

Solanelles et al. [11] designed an electronic control system for pesticide application proportional to the canopy width of tree crops (**Figure 4**). A prototype of an electronic control system based on ultrasonic sensors and proportional solenoid valves for a proportional application to the canopy width of tree crops was mounted on an air-assisted sprayer. The sprayer flow rate adjustment was based on the relationship between the actual tree width measured by the ultrasonic sensors and the maximum tree width of the orchard. The prototype was tested in olive, pear, and apple orchards to assess the system's performance in different crop geometries. Metal tracers were used so that spray deposits for each treatment could be measured on the same samples, reducing sampling variability. Liquid savings of 70, 28, and 39% in comparison to a conventional application were recorded in the olive, pear, and apple orchards, respectively, which resulted in lower spray deposits on the canopy but a higher ratio between the total spray deposit and the liquid sprayer output. A reduction of the maximum tree width parameter in the control algorithm in the apple orchard reduced spray savings but increased spray deposition, with spray savings mainly in the middle level of the outside canopy, compared to conventional air-assisted applications. As a result of this work, the prototype was assembled with ultrasonic sensors with a working range of 0.4–3.0 m.

Gil et al. [27] designed, implemented, and validated a variable-rate sprayer vineyard prototype (**Figure 5**). This prototype can modify the sprayed volume application rate according to the target geometry by using an algorithm based on the canopy volume inspired by the tree row volume model. Variations in canopy width along the row crop are electronically measured

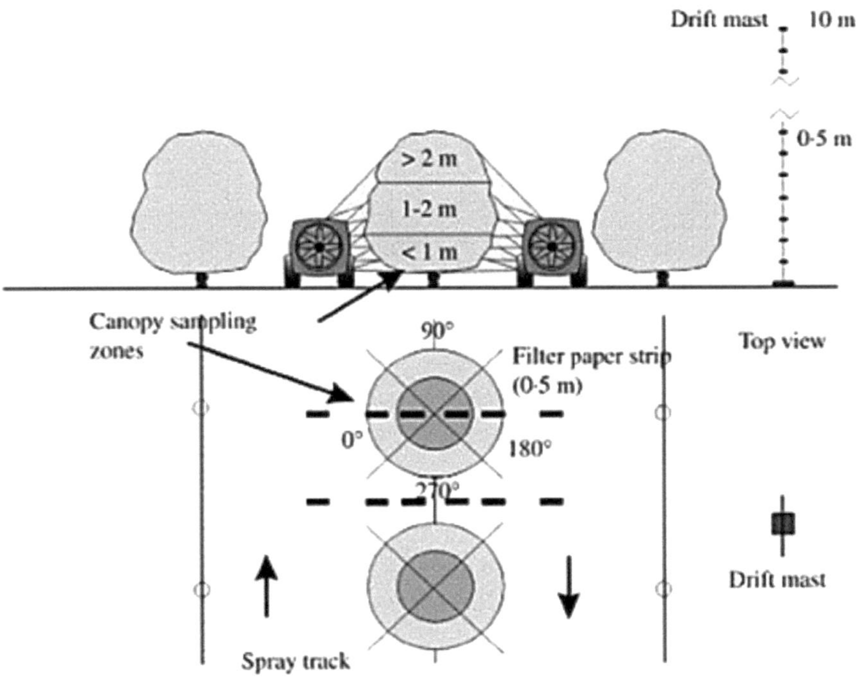

Figure 4. Sampling strategy for one replication in the olive orchard trial [11].

(a) (b) (c) (d)

Figure 5. (a) and (b) Placement of components on the sprayer, (c) laptop for wireless control of the prototype from the tractor cab, (d) interface for input data created using LabVIEW [27].

using several ultrasonic sensors placed on the sprayer and used to modify the emitted flow rate from the nozzles in real time; the objective during this process is to maintain the sprayed volume per unit canopy volume. Field trials carried out at different crop stages for Merlot and Cabernet Sauvignon vines (*Vitis vinifera*) indicated a good relationship between the applied volume and canopy characteristics. The potential pesticide savings were estimated to be 21.9% relative to the costs of a conventional application. This conclusion is in accordance with the results of similar research on automated spraying systems.

Zaman and Salyani [24] evaluated the repeatability of ultrasonic measurements of tree volume, determined the effects of ground speed and foliage density on the ultrasonic measurements, and quantified the difference between volumes of the North and South canopy halves of citrus trees. An experiment was conducted to examine the effects of the canopy foliage density and ground speed on the performance of the Durand-Wayland ultrasonic system in tree volume measurement (**Figure 6**). The difference between ultrasonic and manual volumes ranged from −17.3 to 28.71% at the 95% confidence level. About 95% of the ultrasonic measurements were repeatable within −12.7 to 30.9% of the manual volume. Canopy foliage density had significant effect on ultrasonic measurements of canopy volume. The volume difference was higher in light than dense trees. There was no significant effect of ground speed (1.6–4.7 km/h) on ultrasonic volume measurements. Variability of the measurements in partially defoliated canopies increased as ground speed increased. There was a significant difference between the volumes of two sides of the trees.

Schumann and Zaman [16] developed a software for real-time ultrasonic mapping of tree canopy size. A schematic layout of ultrasonic transducer system and manually measured tree dimensions were used for calculation of tree canopy sizes in a citrus grove. Vehicle and trailer with vertical array of 10 ultrasonic transducers and differential global positioning system (DGPS) were used to measure tree heights and volumes. Transducers are mounted from 0.6 to 6.0 m above the ground (**Figure 7**). The data collected with this automated system were compared with manually measured size data of 30 trees to estimate accuracy, and a grove of 376 citrus trees was surveyed twice with the system to estimate repeatability. Results showed no significant differences between ultrasonically and manually measured tree sizes ranging in height from 2.1 to 4.3 m and in volume from 6.3 to 54.0 $m^3/tree^{-1}$. The system located tree positions for GIS mapping purposes within 1.37 m, 95% of the time.

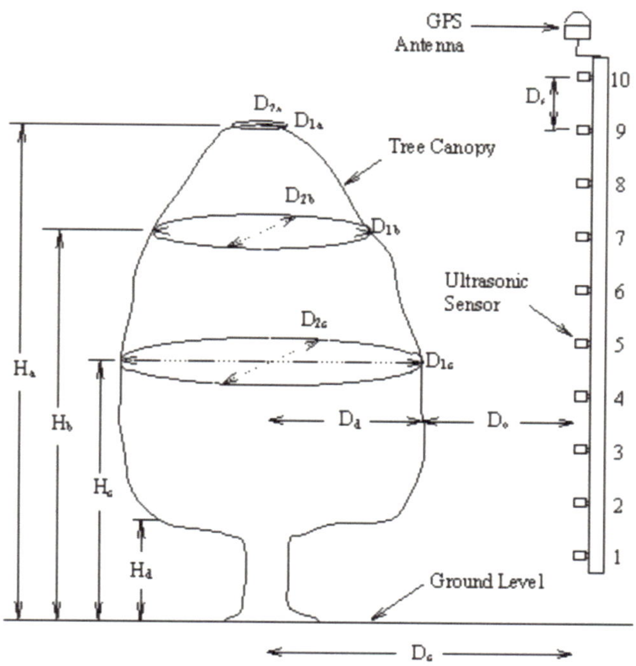

Figure 6. Schematic view of dimensions used to compute canopy volume manually and with ultrasonic measurements [24].

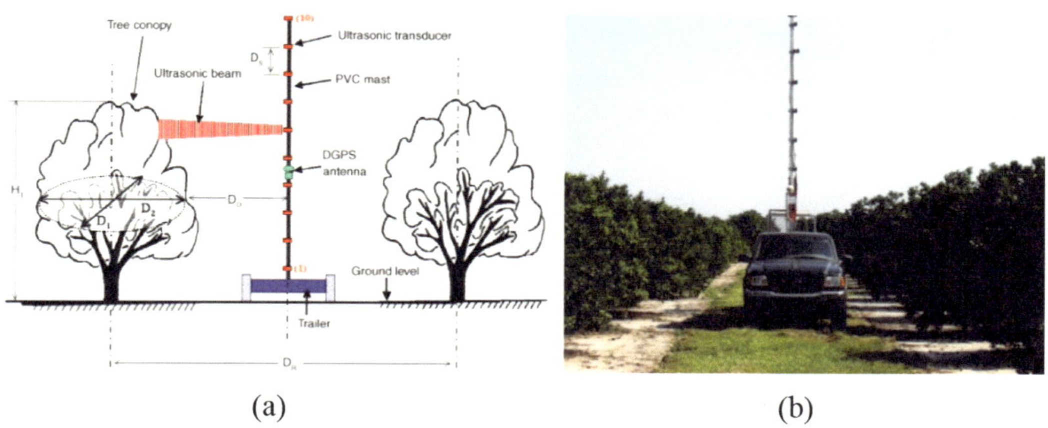

Figure 7. (a) Schematic layout of ultrasonic transducer system, (b) vehicle and trailer with vertical array of 10 ultrasonic transducers.

Palleja and Landers [26] proposed a real-time method, based on an array of ultrasonic sensors, to estimate canopy density in apple orchards and vineyards (**Figure 8**). This estimation could be used as a reference to adjust the canopy spraying machine parameters with the aim of improving deposition and avoiding drift. Two sets of experiments were carried out: the first one using a single ultrasound sensor in a greenhouse to determine the signal behavior and adjust the algorithms. The second set of experiments was conducted in the orchard and vineyard, under real working conditions. Results show that the signal obtained is highly correlated with the growing season, and it has similar values on both sides of the row, with an error of 14.1% in vineyards and 3.8% in apple trees and it is sensitive enough to detect hailstorm effects on the canopy. The ultrasound echoes and the canopy density are proportional. The greater the density, the more the echoes produced. The sprayers incorporate a set of four ultrasound sensors and a Louvre system, which allows air volume to be adjusted from 0 to 100%. Four ultrasound sensors were attached on the front of the sprayer, at 2.2 m from the nozzles, and distributed at different heights.

Palleja and Landers [28] developed a nonexpensive system to estimate the crop density using ultrasound sensors (**Figure 9**). It is important to note that canopy spraying is rarely, if ever, conducted after harvest and it is often done before blossom, in the dormant period. The real-time capabilities of the ultrasonic system allow the sprayer to be adjusted in order to improve spray deposition and reduce spray drift. As well as density, dead plants or row ends are easily detectable, and the sprayer can automatically switch the nozzles on/off.

Maghsoudi et al. [29] designed an electronic control system for the detection and estimation of tree canopy dimensions for application rate adjustment. Three ultrasonic ranging sensors were utilized to estimate the distance to the target at three different heights (**Figure 10**). A multilayer

(a) (b) (c)

Figure 8. (a) Schematic hypothesis diagram, (b) and (c) modified sprayer and ultrasound sensor distribution [26].

(a) (b) (c) (d) (e)

Figure 9. (a–d) Canopy evolution along the season, (e) ditch and tractor tilt problem [28].

Figure 10. (a) Variable-rate sprayer for fruit tree protection, (b) attached nozzles mounted on the vertical masts for orchard tree spraying [29].

perceptron (MLP) neural network with gradient-descent back-propagation algorithm, tangent-sigmoid transfer function, and 3-7-6 topology was used for volume estimation of tree sections. Training and validation errors as well as R^2 values indicated the reliability of the network for volume prediction. Results of t-test for comparing the number of spray droplet impacts, coverage of (artificial) target, spray quality parameter, and relative span factor between variable rate and conventional spraying were not significant, which indicates the consistency of spray distribution in selective application. Experiments showed a reduction in pesticide usage of about 34.5% by means of variable-rate technology (41.3, 25.6, and 36.5, respectively for the top, middle, and bottom sections of tree canopy). Precise application of agrochemicals reduces both costs and environmental pollution by supporting a decrease in the amount of delivered spray.

Jeon et al. [30] evaluated ultrasonic sensor for variable-rate spray applications. Ultrasonic sensors were subjected to simulated environmental (**Figure 11**) and operating conditions to determine their durability and accuracy. Conditions tested included exposure to extended cold, outdoor temperatures, crosswinds, temperature change, dust clouds, travel speeds, and spray cloud effects. After exposure to outdoor cold conditions for 4 months, the root mean square (RMS) error in distance measured by the ultrasonic sensor increased from 3.31 to 3.55 cm, which was not statistically significant. Neither the presence of dust cloud nor the changes in crosswind speeds over a range from 1.5 to 7.5 m/s had significant effects on the mean RMS errors. Varying sensor travel speed from 0.8 to 3.0 m/s had no significant influence on sensor detection distances. Increasing ambient temperature from 16.7 to 41.6°C reduced the detection distance by 5.0 cm. The physical location of the spray nozzle concerning the ultrasonic sensor had a significant effect on mean RMS errors. The mean RMS errors of sensor distance measurements ranged from 2.3 to 83.0 cm. The RMS errors could be reduced to acceptable values by proper controlling of the sensor/spray nozzle spacing on a sprayer. Also, multiple-synchronized sensors were tested for their measurement stability and accuracy (due to possible cross-talk errors) when mounted on a prototype sprayer. It was found that isolating the pathway of the ultrasonic wave of each sensor reduced detecting interference between sensors during multiple sensor operations.

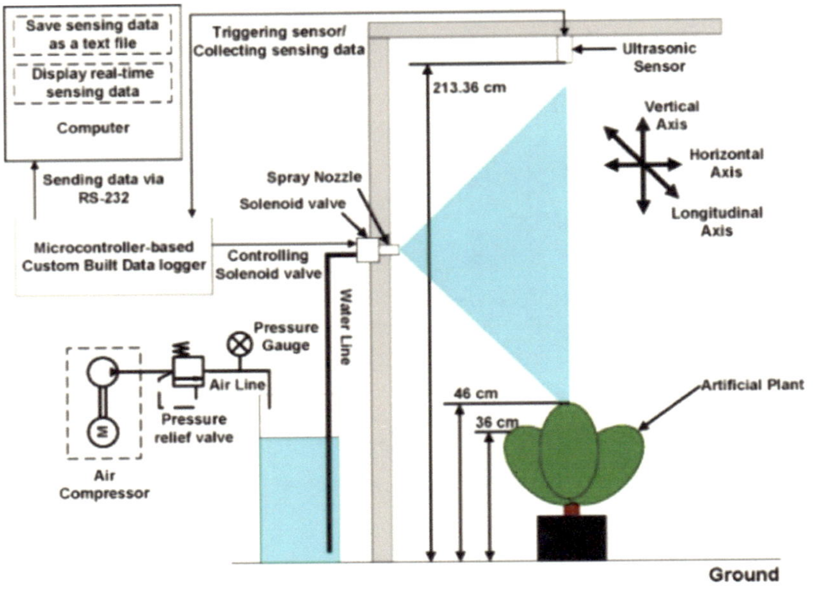

Figure 11. Experiment setup to test the sensor stability with the spray clouds [30].

Of the various types of sensors used in current precision spray systems, ultrasonic sensors that are affordable, relatively robust during outdoor conditions, and capable of estimating the canopy volume of trees satisfactorily have been used by several researchers. It was proved that the ultrasonic system is capable of sensing density. However, it has strong and weak points. The main advantages of ultrasonic sensors are their robustness and low price. Ultrasonic sensors have relatively low costs and can be easily implemented. The system works in real time, and it works through netted canopies, has a small error during most of the season, and can be used as a reference for canopy density. However, the main drawback is the large angle of divergence of ultrasonic waves, and it has to be calibrated and very uneven fields generate inconsistent data. The error remained low up to and including harvest date at the end of September, but significant errors must be expected at the late season. This limits the resolution and accuracy of the measurements taken and also requires the use of many units to cover a typical agricultural scene [31]. The reflection of the sound waves emitted by an ultrasonic sensor is significantly affected by the directional angle and material of the measured plane. Different leaves of a fruit tree have different angles, which will also change when the wind blows. As a result, the angle of tree leaves can easily affect the measurement of the leaf wall area and cause errors in the determination of the distance from the fruit tree and the leaf wall area [32].

4. LIDAR sensor-based detection technology

Another detection principle, which is being used rapidly, is based on the light detection and ranging (LIDAR) sensor technology. This technology is a nondestructive remote sensing technique for the measurement of distances. It is ideal for detecting and measuring nonmetallic or biological objects [33], which provides a relatively novel tool for generating a unique and comprehensive mathematical description of the tree structure. LIDAR is a remote laser range sensor based on the measurement of the elapsed time between the transmission of a pulsed laser beam and the reception of its echo from a reflecting object; this time-of-flight (TOF) is used to estimate the distance between the laser and the object. The advantage of the laser light relative to the ultrasonic waves is that the measurement beam is thinner and less divergent and can be combined with a scanning mechanism to obtain a bidimensional scan pattern to report information about a large area [34]. Terrestrial LIDAR is now used in characterizing canopy structure for different applications like forestry or agriculture.

The use of terrestrial LIDARs in agriculture enables the measurement of structural parameters of the orchards such as the volume of the trees. The ability to very quickly (thousands of points per second) measure the distance between the sensor and the objects around it allows us to obtain 3D cloud points that, by applying appropriate algorithms, makes it possible to digitally reconstruct and describe the structure of trees with high precision [35]. For these reasons, despite their limitation for dusty environments, LIDAR systems have turned out to be one of the most used sensors for the geometric characterization of tree crops.

The capacity of LIDAR to quantify spatial variations, which is an essential aspect of vegetation structure, is a significant advance over some previous methods. LIDAR systems can be used to quantify changes in canopy structure at various time scales, which can provide detailed assessments of canopy growth and allocation responses to field experiments. Laser technology offers unique options regarding the viewing angle and distance information needed to model canopy structure; hence, there is an emergency to thoroughly investigate LIDAR structural applications [36]. The LIDAR system developed made it possible to obtain 3D digitalized images of crops from which a significant amount of plant information, such as height, width, volume, leaf area index, and leaf area density, could be obtained.

In agricultural applications, it is, however, possible to use two-dimensional (2D) terrestrial LIDAR sensors, which are much cheaper to use [37]. 2D LIDAR sensors obtain a point cloud corresponding to a plane or section of the object of interest. The fact that these sensors only scan in one plane does not necessarily limit their scope to 2D perception [38]. Hence, this sensor gives as output a point cloud that, postprocessed, can be exploited for the construction of a 3D image. Rosell Polo et al. [39] proposed the use of a 2D LIDAR scanner in agriculture to obtain 3D structural characteristics of plants. The results obtained for fruit orchards, citrus orchards, and vineyards showed that this technique could provide fast, reliable, and nondestructive estimates of 3D crop structure. It can be concluded that LIDAR systems were able to measure the geometric characteristics of plants with sufficient precision for most agriculture applications.

Early works were concentrated on comparing of manual volume estimation with LIDAR and ultrasonic sensor measurements [23]. Results indicated good correlation between the estimation made by LIDAR and ultrasonic sensors, while correlation with manual measurements was lower. Observation showed larger differences between manual and sensor estimations in less dense trees. This canopy information was used to adjust agrochemical dose rate [40] and estimate fruit yield in citrus groves [41]. LIDAR sensor in relation to vertical sampling resolution can gather much more information from canopy parameters for a more accurate estimation in comparison with array of ultrasonic sensors [26–28, 42]. The results of these tests were satisfactory, but extrapolation of these results to trees with different structures is not easy. Although several groups have developed prototypes to adjust the application flow rate to the variations in canopy structural parameters using ultrasonic sensors, a review of various targeted spraying methods [43] showed that solutions for variable-rate spraying in orchards are still in prototype phase; however, there are already commercially available sprayers for weed control and plant fertilization in arable land.

Rosell et al. [44] proposed a method of 2D LIDAR scanner in agriculture to obtain three-dimensional (3D) structural characteristics of plants (**Figure 12**). There was a great degree of concordance between the physical dimensions, shape, and global appearance of the 3D digital plant structure and the real plants, revealing the coherence of the 3D tree model obtained from the developed system with respect to the real structure. For some selected trees, the correlation coefficient obtained between manually measured volumes and those obtained from the 3D LIDAR models was as high as 0.976.

Escolà et al. [13] designed, implemented, and validated a prototype (**Figure 13**) running a variable-rate algorithm to adapt the volume application rate to the canopy volume in orchards

(a) (b) (c) (d)

Figure 12. The LIDAR measurement system, (a) data in Cartesian coordinates, (b) data in polar coordinates, (c) pear orchard, (d) different views of the 3D structure [44].

Figure 13. Variable-rate orchard sprayer prototype implemented with LIDAR sensor [13].

on a real-time and continuous basis. The orchard prototype was divided into three parts: the canopy characterization system (using a LIDAR sensor), the controller executing a variable-rate algorithm, and the actuators. The controller determines the intended flow rate by using an application coefficient (required liquid volume per unit canopy volume) to convert canopy volume into a flow rate. The sprayed flow rates are adjusted via electromagnetic variable-rate valves. The goal of the prototype was to keep the actual application coefficients as close as possible to the objective. Strong relationships were observed between the intended and the sprayed flow rates ($R^2 = 0.935$) and between the canopy cross-sectional areas and the sprayed flow rates ($R^2 = 0.926$). In addition, when spraying in variable-rate mode, the prototype achieved significantly closer application coefficient values to the objective than those obtained in conventional spraying application mode.

Palleja and Landers [28] analyzed the sensitivity of the tree volume estimates in the spatial trajectory of a LIDAR (**Figure 14**) relative to different error sources. The sequence of two-dimensional scans performed with a LIDAR attached to a tractor can be interpreted as the three-dimensional outline of the trees of the grove and used to estimate their volume. The sensitivity of the tree volume estimates relative to different error sources in the estimated spatial trajectory of the LIDAR is analyzed. Tests with pear trees have demonstrated that the estimation of the volume is very sensitive to errors in the determination of the distance from the LIDAR to the center of the trees (with errors up to 30% for an error of 50 mm) and in the determination of the angle of orientation of the LIDAR (with errors up to 30% for misalignments of 2°). Therefore, any experimental procedure for tree volume estimate based on a motorized terrestrial LIDAR scanner must include additional devices or procedures to control or estimate and correct these error sources.

The main advantages of LIDAR sensors are their high speed and accuracy of measurement, and they provide a 3D point cloud of the object being measured. LIDAR sensors facilitate the

Figure 14. LIDAR placed on the back of a tractor [28].

description of the geometric structure of trees. However, the scale of these remote sensing techniques is relatively large, and consequently, the sensing resolution may be insufficient for a real-time variable-rate application in a liner production field. In addition, remote sensing techniques typically have a chronological gap between detection and application, resulting in application errors. To reduce this problem, a LIDAR system or a laser scanner has been used to measure canopy volume. Promising results were reported for using this system in which measured canopy volume was close to manually measured volume [39, 45, 46]. Unfortunately, the narrow row spacing in a liner field may restrict LIDAR from being used on variable-rate liner sprayers. It is also a relatively expensive sensor ($2000–6000), and the high cost of these instruments limits their use. Furthermore, a typical tree liner sprayer treats multiple rows at a time. Each liner row would require an individual LIDAR system to measure its tree canopy variation. Thus, controlling a variable-rate application sprayer would require several LIDAR systems. This would increase the application cost to an impractical level.

5. Computer stereo vision-based detection technology

A video camera can capture video images of fruit trees and segregate parameters such as the leaf wall area, height, and density based on the color information through video processing techniques. However, due to a lack of measured distance information, distance can only be estimated based on the precalibrated distance from the video camera, which may easily generate relatively large errors. Computer stereo vision implies the extraction of 3D information from digital images, as obtained by a CCD or CMOS image sensor-based digital camera, which can provide a 3D field image by combining two monocular field images taken simultaneously using a binocular camera [47]. The main advantage of stereoscopic vision over conventional monocular vision is its ability to detect ranges: distances between scene objects and the camera. Monocular cameras create planar images in which each pixel is the result of a two-dimensional (2D) projection of the 3D world. Stereovision adds a third coordinate, or range, which completes the full localization of any point within a 3D Cartesian frame. The natural outcome of a stereovision sensor is a 3D point cloud that renders the captured scene with a degree of detail proportional to the resolution of the acquired images. Every single point in the 3D cloud comes from a stereo-matched pixel and will be endowed with three coordinates that identify its exact spatial position [38].

Berenstein et al. [48] proposed grape clusters and foliage detection algorithms for autonomous selective vineyard sprayer (**Figure 15**). Novel machine vision algorithms were developed to detect gaps between grapevines in order to reduce pesticide use during foliage spraying and to detect the exact location of grape clusters to target spraying toward them. A spraying robot equipped with these detection capabilities and a pan/tilt head with a spray nozzle would be able to spray selectively and precisely, reducing significant amounts of spraying material and human labor. Results show 90% accuracy of grape cluster detection leading to 30% reduction in the use of pesticides.

Microsoft's Kinect system can capture the color and depth information of a scene in real time. This system consists of a red, green, and blue (RGB) video camera, a monochrome complementary metal-oxide semiconductor (CMOS) video camera, and an IR transmitter. The color CMOS camera generates color images, and the IR transmitter and the IR CMOS camera generate depth images. The Kinect system outputs a 640 × 480 RGB image and an IR depth image. Because conventional depth sensors (e.g., laser ranging radars) are deficient concerning sensitive information readability, depth cameras have become an essential means for measuring the depth-of-field information of scenes. Under ideal conditions, the resolution of depth information acquired by a depth camera can reach 3 mm. Xiao et al. [32] designed an intelligent precision orchard pesticide spray technique based on the depth-of-field extraction algorithm (**Figure 16**). To obtain desirable spray effect, the advantages of color and depth information using Microsoft's Kinect system were integrated. To adjust and control the spray intensity of sprayers and the dose of sprayed pesticides, an equation for calculating the leaf wall area average distance of fruit trees was proposed. A comparison with the measured distances showed that the distances calculated

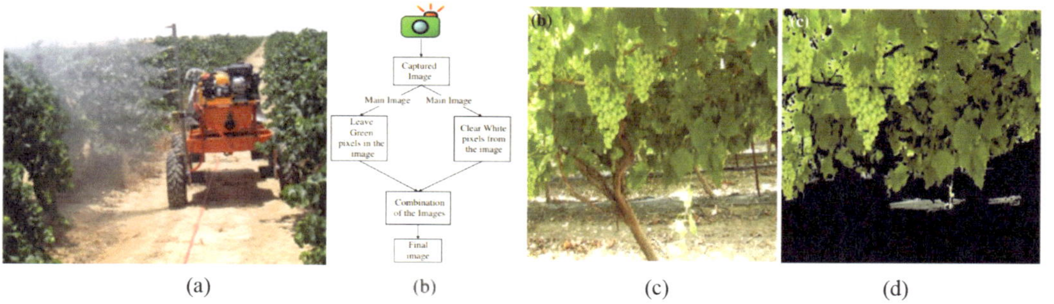

Figure 15. (a) Vineyard spraying robot, (b) block diagram of the algorithm, (c) captured image, (d) final foliage image.

Figure 16. The procedure of target tree extraction. (a) Color image, (b) segmented image, (c) depth image, (d) 3D layer in-depth image, (e) comparative image, (f) resultant image.

based on the data acquired by the Kinect system were accurate. The results of the experiment on peach trees, apricot trees, and grapevines demonstrated that the intelligent orchard pesticide precision spray model established based on the average distance and the leaf wall area density can improve the efficiency in spraying pesticides, reduce waste and environmental pollution, and achieve automated and precision orchard production.

6. Advanced data fusion application technique and future directions

The integration of data and knowledge from several sources is known as data fusion. To overcome the inherent drawbacks and combine the advantages of different kinds of sensors, multimodal sensor fusion has been widely used [49–51]. Briefly, data fusion can be defined as a combination of multiple sources to obtain improved information; in this context, improved information means less expensive, higher quality, or more relevant information. Data fusion is the process of integrating multiple data sources to produce more consistent, accurate, and useful information than that provided by any individual data source. Fusion of the data from two sources (dimensions 1 and 2) can yield a classifier superior to any classifiers based on dimension 1 or dimension 2 alone [52]. In general, all tasks that demand any parameter estimation from multiple sources can benefit from the use of data/information fusion methods. Data fusion techniques have been extensively employed on multisensory environments with the aim of fusing and aggregating data from different sensors. The goal of using data fusion in multisensory environments is to obtain a lower detection error probability and a higher reliability by using data from multiple distributed sources [53].

The use of spatial sensors with the agricultural application has increased rapidly in recent years as their costs decline. Because of their ability to provide instantaneous information that can be used for feature extraction and object detection, vision systems and laser scanners are becoming more common in outdoor agricultural applications such as tree detection, map construction, mobile robot localization, and navigation. Vision systems are low-cost solutions for extracting different features (e.g., color, edge, and texture), while laser scanners are popular sensors in outdoor applications as they provide precise range and angle measurements in large angular fields. Fusing images from cameras with range data from laser scanners enable mobile robots and vehicles to more confidently perform a variety of tasks in outdoor environments [49]. There are differences between the data acquired from the laser scanner and the camera images. The 2D laser scanner generates a single horizontal scan of the environment, whereas the camera provides an instantaneous image of the local environment with precise depth information. A laser scanner provides range and bearing data, while the camera primarily provides intensity and color information. There are some standard features in both types of data. For example, many corners and edges correspond to a sudden change in the range of the laser scan data and a sudden variation in image intensity [54].

Shalal et al. [55, 56] presented a novel tree trunk detection algorithm using camera and laser scanner data fusion (**Figure 17**). The innovation and contribution of this study developed a new tree trunk detection algorithm using low-cost camera and laser scanner data fusion as a

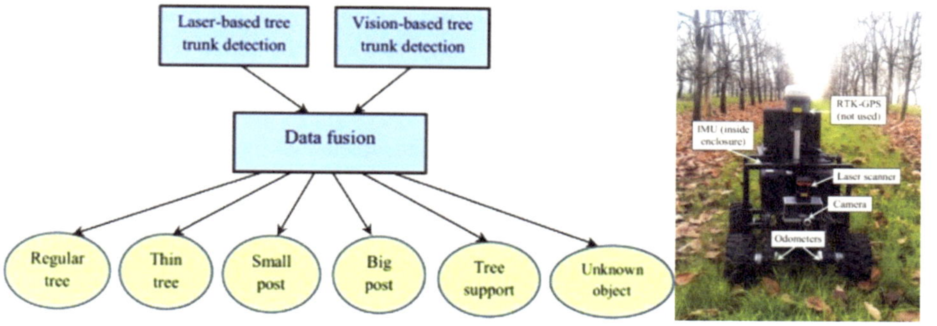

(a) (b)

Figure 17. (a) The block diagram of the two tree trunk detection algorithms, (b) explorer platform with onboard sensors.

component of fully automated operation to enhance the detection capability and to discriminate between trees and nontree objects. The laser scanner is used to detect the edge points and determine the width of the tree trunks and nontree objects, while the camera images are used to verify the color and the parallel edges of the tree trunks and nontree objects. The algorithm automatically adjusts the color detection parameters after each test, which shows to increase the detection accuracy. The algorithm was able to detect the tree trunks and discriminate between trees and nontree objects with a detection accuracy of 96.64% showing that the fusion of both vision and laser scanner technologies produced robust tree trunk detection. Fusion of data from these sensors was found to improve tree detection because the laser scanner can provide reliable ranges, angles, and width of the tree trunks and nontree objects, while the vision system can distinguish between tree trunks and other nontree objects from different features.

Data fusion as a new method was demonstrated for detecting trees and nontree objects using a camera and laser scanner data fusion. The utilization of both camera and laser scanner data enhanced the tree trunk detection. Projecting from the laser scanner to the image plane and selecting the region of interest with the required features were useful since it reduces the processing time and minimizes the effect of the noise in the other parts of the image. The developed algorithm relies only on the onboard sensors without adding any artificial landmarks such as tags or reflective tapes on the trees in the orchard. The algorithm automatically adjusts the color detection parameters after each test, which was observed to improve the detection accuracy. Above all, the fusion of data from the vision and laser sensors improves plant canopy detection because the laser scanner can provide accurate ranges, angles, and widths of the tree and objects, while the vision system can distinguish between a tree and other objects.

7. Discussion and conclusion

Development of new, environmentally friendly alternative variable-rate sprayer application techniques only began in the last four decades. Its objective has been to use variable-rate sprayer dosage rates that are as low as possible and to apply variable-rate sprayer only to places where this was necessary, with minimum losses transferred to the environment. Various procedures and methods for tree canopy detection have been suggested and developed by both computer and agricultural scientists [57]. The detailed review indicates that the establishment of an appropriate variable-rate sprayer is still one of the critical issues in plant protection. Improvement of electronic tree canopy sensing should facilitate electronic measurements of the tree canopy characteristics and enable more precise control of variable-rate sprayer dosage, which can then ensure a faster response of the entire system at higher driving speeds in the orchard. Some researchers suggested that electronic characterization of the tree canopy could be carried out more efficiently by using some detection approaches, including ultrasonic, imaging, and optical detection systems.

The analysis of sensing systems for electronic canopy characterization indicates that the infrared and ultrasonic sensors as the oldest and simplest approaches are still an appropriate tool for determining average canopy characteristics such as the ends of rows and significant gaps between well-separated trees. Furthermore, when equipped with appropriate software, the infrared and ultrasonic sensor transceivers can be used for measuring the tree density. For this reason, these types of sensors will remain on sprayers in the near future, because it can simplify the operator's repetitive work in the orchards and might serve as an input parameter for adjusting variable-rate sprayer dosage from a particular nozzle.

The analysis of the different existing detection systems to characterize the structure of tree plantations shows the existence of several aspects that limit the use of most of the systems under field conditions, some sensors remaining suitable for this purpose. Laser scanners and stereo vision are direct competitors and are probably the most promising and complementary techniques for achieving 3D maps of plants and canopies, although infrared and ultrasonic sensors remain an attractive option for specific applications. In fact, the possibilities of combining sensors for this purpose are innumerable. In the near future, it is highly likely that we will see a notable advance in this field of research with the increased use of the new generation of flash LIDAR sensors, capable of measuring 3D structures of plants in real time and at a moderate cost compared to alternative detection systems.

The usefulness of using camera sensor to facilitate the quantification of the density of the plantations has also been mentioned. However, it has become clear that there is still a long way to go and both the geometric characterization of plants and variable application techniques must be improved. More highlighted advanced stereo vision measurement sensing systems for electronic canopy characterization sound very attractive for detection of the tree canopy, because this technique captures a massive image of an orchard in a short time. However, the computer-generated digital 3D terrain model of the orchard still cannot assure characterization of canopy diameter, height, and number of leaves with sufficient precision for estimating the leaf area index needed for appropriate adjustment of the variable-rate sprayer dosage.

In this chapter, variable-rate sprayer applications based on real-time sensor technologies have been reviewed. Based on the results from reports and literatures, **Table 1** summarizes the operating principles and the main pros and cons of the exposed sensors and methods for the measurement of the geometrical properties of plants and crops.

With regard to agricultural applications, innovative techniques represent an essential contribution to the improvement of variable-rate sprayer application. The different sensing system can detect tree canopy characteristics precisely, and when combined with sophisticated decision-making

Sensors	Measuring principle	Pros	Cons
Infrared sensors	All objects with a temperature above absolute zero emit heat energy in the form of radiation. Infrared sensors measure infrared light radiating from objects in their field of view. Work entirely by detecting infrared radiation emitted by or reflected from objects.	Temperature and humidity have little impact on the detection results. Measurement relatively independent of atmospheric conditions.	Accurate measurement of the 3D characteristics of the canopy remains unfeasible for the moment. Plant appearance, light intensity, walking speed, and plant space have evident influences on detecting effect. Deficient spatial resolution for applications in agriculture. Short detectable distance, complicated circuit.
Ultrasonic sensors	Measure the distance to an object by using sound waves. Based on determining the flight time of an ultrasonic wave from the point of emission to the point of detection after bouncing off an object.	Robustness and low price make ultrasonic sensors suitable for agricultural applications. Relatively easy to implement.	The large angle of divergence of ultrasonic wave beams limits the resolution and accuracy of the measurements taken. The use of many units to cover a common agricultural scene is required.
LIDAR sensors	Based on the measurement of the distance from a laser emitter to an object or surface using a pulsed laser beam. Time-of-flight LIDAR measures the time that a laser pulse takes to travel between the sensor and the target.	High speed of measurement allows obtaining cloud points quickly. Applying appropriate algorithms makes it possible to digitally reconstruct and describe the structure of trees with high precision. Plant information, such as height, width, volume, leaf area index, and leaf area density can be obtained with sufficient precision.	The estimation of the volume is very sensitive to errors in the determination of the distance from the LIDAR to the center of the trees and in the determination of the angle of orientation of the LIDAR. Motorized terrestrial LIDAR scanners must include additional devices or procedures to control or estimate and correct these error sources.
Stereo vision	Provides a 3D field image by combining two monocular field images taken simultaneously using a binocular digital camera. Computer algorithms are necessary to convert the original camera coordinate arrays of the objects into their real-world coordinates.	Provides realistic 3D image of plants and tree crops. Measures directly the 3D vegetation structure including those plant physical parameters that are important for production management, such as crop size and volume.	Offer less accuracy than laser-based systems and need appropriate calibration and recording procedures. When several images are processed together, the magnitude of the data files grows considerably, complicating the handling and storage of 3D information and requiring long processing times. The problem becomes more critical when real-time processing is required.

Table 1. Physical principles and most remarkable characteristics of the main systems used for the geometrical characterization of tree crops and their main advantages and disadvantages.

models, they enable accurate variable-rate sprayer dosage control. The coordinated use of multiple sensors, the development of new real-time data processing algorithms, and the simplification of crop adaptable application systems are objectives for the future of this research line. Obtaining a precise geometrical characterization of a crop at any point during its production cycle by means of a new generation of affordable and easy-to-use detection systems, such as LIDAR and stereo vision systems, will help to establish precise estimations and provide valuable information on which to base more sustainable pesticide dosages. Without any doubt, optical sensing systems for electronic canopy characterization including a LIDAR sensor provide the most accurate and detailed information about the tree canopy. When supported with the proper software, a LIDAR-based signal can represent a perfect tool for creating a 3D space at low installation costs, which is essential for guiding a robotic arm equipped with nozzles and small vents along the tree row in real time. For all these reasons, LIDAR will represent the crucial sensor in the further development of both trailer-mounted and self-propelled sprayer prototypes, which should find the widespread commercial application.

In the near future, the evolution and development of new sensors devoted to the geometric characterization of tree crops will enable significant and much needed advances in optimizing the use of variable-rate sprayer in agriculture, as well as an increase in production and quality by improving training systems. It is worth noting that the benefits of variable spray affect millions of cultivated hectares and therefore impact directly on the society and the environment in which we live. It is therefore of vital importance to continue devoting major efforts to the development of increasingly accurate, robust, and affordable systems capable of measuring the geometric characteristics of plantations, which support the development of the different areas of a sustainable and precision agriculture. However, it is still necessary to resolve several technological and commercial questions. The former include improving detection systems, especially with regard to developing software for the postprocessing steps and improving the speed of calculation and decision making. Among the latter, it is essential to produce low-cost sensors and control systems to facilitate large-scale deployment.

Acknowledgements

This research project was financially supported by the National Natural Science Foundation of China for Young Scholars (Grant No. 51605210), by Yunnan Applied Basic Research Youth Project (Grant No. 2015FD011), by Introduced Talents Scientific Research Staring Foundation of KMUST (Grant No. 14118940), by Scientific Research Fund Project of Yunnan Provincial Education Department (Grant No. 2015Y079), and by Opening Project of the Key Laboratory of Bionic Engineering (Ministry of Education) (K201621), Jilin University.

Author details

Zhihong Zhang[1,2]*, Xiaoyang Wang[1], Qinghui Lai[1] and Zhaoguo Zhang[1]

*Address all correspondence to: zzh_0822@hotmail.com

1 Faculty of Modern Agricultural Engineering, Kunming University of Science and Technology, Kunming, China

2 United States Department of Agriculture, Agricultural Research Service, Wooster, OH, USA

References

[1] Ladd Jr T, Reichard D, Collins D, et al. An automatic intermittent sprayer: A new approach to the insecticidal control of horticultural insect pests 1 2. Journal of Economic Entomology. 1978;**71**(5):789-792

[2] Giles D, Delwiche M, Dodd R. Control of orchard spraying based on electronic sensing of target characteristics. Transactions of the ASAE. 1987;**30**(6):1624-1636

[3] Steward BL, Tian FL, Tang L. Distance-based control system for machine vision-based selective spraying. Transactions of the ASAE. 2002;**45**(5):1255. DOI: https://doi.org/10.13031/2013.11053

[4] Tian L, Reid JF, Hummel JW. Development of a precision sprayer for site-specific weed management. Transactions of the ASAE. 1999;**42**(4):893. DOI: https://doi.org/10.13031/2013.13269

[5] Moltó E, Martín B, Gutiérrez A. Pm—Power and machinery: Pesticide loss reduction by automatic adaptation of spraying on globular trees. Journal of Agricultural Engineering Research. 2001;**78**(1):35-41. DOI: https://doi.org/10.1006/jaer.2000.0622

[6] He X, Zeng A, Liu Y, et al. Precision orchard sprayer based on automatically infrared target detecting and electrostatic spraying techniques. The International Journal of Agricultural and Biological Engineering. 2011;**4**(1):35-40

[7] Von-Bargen K, Meyer GE, Mortensen DA, et al. Red/near-infrared reflectance sensor system for detecting plants. Optics in Agriculture and Forestry. International Society for Optics and Photonics. 1993;**1836**:231-239. http://dx.doi.org/10.1117/12.144032

[8] Jiao J, Chu J-Y, Ma G-B. Application of infrared photoelectrics switch for sprarying on aspen. Journal of Agricultural Mechanization Research. 2005;**2005**(3):216-217

[9] Jianjun Z, Aijun Z, Xiongkui H, et al. Research and development of infrared detection system for automatic target sprayer used in orchard. Transactions of the Chinese Society of Agricultural Engineering. 2007;**23**(1):129-132

[10] Chen Z, Wang Y, Meng T, et al. Experiment of infrared target detection for plants pesticide spraying. Drainage and Irrigation Machinery. 2009;**27**(4):237-246

[11] Solanelles F, Escolà A, Planas S, et al. An electronic control system for pesticide application proportional to the canopy width of tree crops. Biosystems Engineering. 2006;**95**(4):473-481. DOI: https://doi.org/10.1016/j.biosystemseng.2006.08.004

[12] Downey D, Giles D, Klassen P, et al. "Smart" sprayer technology provides environmental and economic benefits in california orchards. California Agriculture. 2011;**65**(2):85-89

[13] Escolà A, Rosell-Polo J, Planas S, et al. Variable rate sprayer. Part 1–orchard prototype: Design, implementation and validation. Computers and Electronics in Agriculture. 2013;**95**:122-135

[14] Planas S, Solanelles F, Fillat A. Pm—Power and machinery: Assessment of recycling tunnel sprayers in mediterranean vineyards and apple orchards. Biosystems Engineering. 2002;**82**(1):45-52. DOI: https://doi.org/10.1006/bioe.2001.0055

[15] Gil E, Escola A, Rosell J, et al. Variable rate application of plant protection products in vineyard using ultrasonic sensors. Crop Protection. 2007;**26**(8):1287-1297

[16] Schumann AW, Zaman QU. Software development for real-time ultrasonic mapping of tree canopy size. Computers and Electronics in Agriculture. 2005;**47**(1):25-40. DOI: https://doi.org/10.1016/j.compag.2004.10.002

[17] Planas S, Rosell JR, Pomar J, et al. Performance of an ultrasonic ranging sensor in apple tree canopies. Sensors. 2011;**11**(3):2459-2477

[18] Moltó E, Martín B, Gutiérrez A. Pm—Power and machinery: Design and testing of an automatic machine for spraying at a constant distance from the tree canopy. Journal of Agricultural Engineering Research. 2000;**77**(4):379-384

[19] McConnell RL, Elliot KC, Blizzard SH, et al. Electronic measurement of tree-row-volume. National Conference on Agricultural Electronics Applications, Hyatt Regency Illinois Center, Chicago, (USA), American Society of Agricultural Engineers; 1983

[20] Giles DK, Delwiche MJ, Dodd RB. Electronic measurement of tree canopy volume. Transactions of the ASAE. 1988;**31**(1):264-272

[21] Brown D, Giles D, Oliver M, et al. Targeted spray technology to reduce pesticide in runoff from dormant orchards. Crop Protection. 2008;**27**(3):545-552

[22] Balsari P, Tamagnone M. An ultrasonic airblast sprayer. International Conference on Agricultural Engineering GBR. 1998:585-586

[23] Tumbo S, Salyani M, Whitney JD, et al. Investigation of laser and ultrasonic ranging sensors for measurements of citrus canopy volume. Applied Engineering in Agriculture. 2002;**18**(3):367

[24] Zaman QU, Salyani M. Effects of foliage density and ground speed on ultrasonic measurement of citrus tree volume. Applied Engineering in Agriculture. 2004;**20**(2):173. DOI: https://doi.org/10.13031/2013.15887

[25] Balsari P, Doruchowski G, Marucco P, et al. A system for adjusting the spray application to the target characteristics. Agricultural Engineering International: CIGR Journal. 2008; **10**:1-12

[26] Palleja T, Landers AJ. Real time canopy density estimation using ultrasonic envelope signals in the orchard and vineyard. Computers and Electronics in Agriculture. 2015;**115**: 108-117

[27] Gil E, Llorens J, Llop J, et al. Variable rate sprayer. Part 2 – Vineyard prototype: Design, implementation, and validation. Computers and Electronics in Agriculture. 2013;**95**(Supplement C):136-150. DOI: https://doi.org/10.1016/j.compag.2013.02.010

[28] Palleja T, Landers AJ. Real time canopy density validation using ultrasonic envelope signals and point quadrat analysis. Computers and Electronics in Agriculture. 2017;**134**(Supplement C):43-50. DOI: https://doi.org/10.1016/j.compag.2017.01.012

[29] Maghsoudi H, Minaei S, Ghobadian B, et al. Ultrasonic sensing of pistachio canopy for low-volume precision spraying. Computers and Electronics in Agriculture. 2015;**112**(Supplement C):149-160. DOI: https://doi.org/10.1016/j.compag.2014.12.015

[30] Jeon HY, Zhu H, Derksen R, et al. Evaluation of ultrasonic sensor for variable-rate spray applications. Computers and Electronics in Agriculture. 2011;**75**(1):213-221. DOI: https://doi.org/10.1016/j.compag.2010.11.007

[31] Rovira-Más F, Zhang Q, Reid J. Creation of three-dimensional crop maps based on aerial stereoimages. Biosystems Engineering. 2005;**90**(3):251-259

[32] Xiao K, Ma Y, Gao G. An intelligent precision orchard pesticide spray technique based on the depth-of-field extraction algorithm. Computers and Electronics in Agriculture. 2017; **133**(Supplement C):30-36. DOI: https://doi.org/10.1016/j.compag.2016.12.002

[33] Bietresato M, Boscariol P, Gasparetto A, et al. On the design of a mechatronic mobile system for laser scanner based crop monitoring. Proceedings of the 14th Mechatronics Forum International Conference, Mechatronics; 2014

[34] Wehr A, Lohr U. Airborne laser scanning—An introduction and overview. ISPRS Journal of Photogrammetry and Remote Sensing. 1999;**54**(2):68-82

[35] Gorte B, Pfeifer N. Structuring laser-scanned trees using 3D mathematical morphology. International Archives of Photogrammetry and Remote Sensing. 2004;**35**(B5):929-933

[36] Van der Zande D, Hoet W, Jonckheere I, et al. Influence of measurement set-up of ground-based lidar for derivation of tree structure. Agricultural and Forest Meteorology. 2006; **141**(2):147-160

[37] Walklate PJ, Cross JV, Richardson GM, et al. It—Information technology and the human interface: Comparison of different spray volume deposition models using lidar measurements of apple orchards. Biosystems Engineering. 2002;**82**(3):253-267. DOI: https://doi.org/10.1006/bioe.2002.0082

[38] Rovira-Más F, Reid J, Zhang Q. Stereovision data processing with 3D density maps for agricultural vehicles. Transactions of the ASABE. 2006;**49**(4):1213-1222

[39] Rosell Polo JR, Sanz Cortiella R, Llorens Calveras J, et al. A tractor-mounted scanning lidar for the non-destructive measurement of vegetative volume and surface area of tree-row plantations: A comparison with conventional destructive measurements. Biosystems Engineering. 2009;**102**(2):128-134

[40] Zaman Q, Schumann A, Miller W. Variable rate nitrogen application in florida citrus based on ultrasonically-sensed tree size. Applied Engineering in Agriculture. 2005;**21**(3): 331-335

[41] Zaman Q, Schumann A, Hostler H. Estimation of citrus fruit yield using ultrasonically-sensed tree size. Applied Engineering in Agriculture. 2006;**22**(1):39-44

[42] Palleja T, Tresanchez M, Teixido M, et al. Sensitivity of tree volume measurement to trajectory errors from a terrestrial lidar scanner. Agricultural and Forest Meteorology. 2010; **150**(11):1420-1427

[43] Van-De-Zande J, Achten V, Michielsen J, et al. Towards More Target Oriented Crop Protection. International advances in pesticide application. 2008;**84**:1245-1252

[44] Rosell JR, Llorens J, Sanz R, et al. Obtaining the three-dimensional structure of tree orchards from remote 2d terrestrial lidar scanning. Agricultural and Forest Meteorology. 2009;**149**(9):1505-1515. DOI: https://doi.org/10.1016/j.agrformet.2009.04.008

[45] Lee K-H, Ehsani R. Comparison of two 2d laser scanners for sensing object distances, shapes, and surface patterns. Computers and Electronics in Agriculture. 2008;**60**(2):250-262

[46] Wei J, Salyani M. Development of a laser scanner for measuring tree canopy characteristics: Phase 2. Foliage density measurement. Transactions of the ASAE. 2005;**48**(4):1595-1601

[47] Kise M, Zhang Q, Rovira Más F. A stereovision-based crop row detection method for tractor-automated guidance. Biosystems Engineering. 2005;**90**(4):357-367. DOI: https://doi.org/10.1016/j.biosystemseng.2004.12.008

[48] Berenstein R, Shahar OB, Shapiro A, et al. Grape clusters and foliage detection algorithms for autonomous selective vineyard sprayer. Intelligent Service Robotics. 2010;**3**(4):233-243. DOI: 10.1007/s11370-010-0078-z

[49] Garcia-Alegre MC, Martin D, Guinea DM, et al. Real-time fusion of visual images and laser data images for safe navigation in outdoor environments. Sensor Fusion-Foundation and Applications. 2011:121-138. InTech. https://doi.org/10.5772/16690

[50] Panich S, Afzulpurkar N. Sensor fusion techniques in navigation application for mobile robot. Sensor Fusion-Foundation and Applications. 2011:101-120. InTech. https://doi.org/10.5772/16444

[51] Thomas C, Narayanaswamy B. Sensor fusion for enhancement in intrusion detection. Sensor Fusion-Foundation and Applications. 2011:61-76. https://doi.org/10.5772/21550

[52] Viola P, Wells III WM. Alignment by maximization of mutual information. International Journal of Computer Vision. 1997;**24**(2):137-154

[53] Castanedo F. A review of data fusion techniques. The Scientific World Journal. 2013;**2013**(19):1-20. DOI: 10.1155/2013/704504

[54] Peynot T, Kassir A. Laser-camera data discrepancies and reliable perception in outdoor robotics. Intelligent Robots and Systems (IROS), IEEE/RSJ International Conference. IEEE. 2010:2625-2632. https://doi.org/10.1109/IROS.2010.5648934

[55] Shalal N, Low T, McCarthy C, et al. Orchard mapping and mobile robot localisation using on-board camera and laser scanner data fusion – Part B: Mapping and localisation.

Computers and Electronics in Agriculture. 2015;**119**(Supplement C):267-278. DOI: https://doi.org/10.1016/j.compag.2015.09.026

[56] Shalal N, Low T, McCarthy C, et al. Orchard mapping and mobile robot localisation using on-board camera and laser scanner data fusion – Part A: Tree detection. Computers and Electronics in Agriculture. 2015;**119**(Supplement C):254-266. DOI: https://doi.org/10.1016/j.compag.2015.09.025

[57] Rosell JR, Sanz R. A review of methods and applications of the geometric characterization of tree crops in agricultural activities. Computers and Electronics in Agriculture. 2012;**81**(Supplement C):124-141. DOI: https://doi.org/10.1016/j.compag.2011.09.007

Chapter 5

Robotic Harvesting of Fruiting Vegetables: A Simulation Approach in V-REP, ROS and MATLAB

Redmond R. Shamshiri, Ibrahim A. Hameed,
Manoj Karkee and Cornelia Weltzien

Additional information is available at the end of the chapter

http://dx.doi.org/10.5772/intechopen.73861

Abstract

In modern agriculture, there is a high demand to move from tedious manual harvesting to a continuously automated operation. This chapter reports on designing a simulation and control platform in V-REP, ROS, and MATLAB for experimenting with sensors and manipulators in robotic harvesting of sweet pepper. The objective was to provide a completely simulated environment for improvement of visual servoing task through easy testing and debugging of control algorithms with zero damage risk to the real robot and to the actual equipment. A simulated workspace, including an exact replica of different robot manipulators, sensing mechanisms, and sweet pepper plant, and fruit system was created in V-REP. Image moment method visual servoing with eye-in-hand configuration was implemented in MATLAB, and was tested on four robotic platforms including Fanuc LR Mate 200iD, NOVABOT, multiple linear actuators, and multiple SCARA arms. Data from simulation experiments were used as inputs of the control algorithm in MATLAB, whose outputs were sent back to the simulated workspace and to the actual robots. ROS was used for exchanging data between the simulated environment and the real workspace via its publish-and-subscribe architecture. Results provided a framework for experimenting with different sensing and acting scenarios, and verified the performance functionality of the simulator.

Keywords: agricultural robots, automated harvesting, simulation, visual servo control, image processing

1. Introduction

Traditional harvesting of fruiting vegetables for fresh market is a labor-intensive task that demands shifting from tedious manual operation to a continuously automated harvesting.

In spite of the advances in agricultural robotics, million tons of fruits and vegetables are still hand-picked every year in open-fields and greenhouses (**Figure 1**). Other than the high labor cost, the availability of the skilled workforce that accepts repetitive tasks in the harsh field conditions impose uncertainties and timeliness costs. For robotic harvesting to be cost-effective, fruit yield needs to be maximized to compensate the additional automation costs. This leads to growing plants at higher densities which make it even harder for an autonomous robot to simultaneously detect the fruit, localize, and harvest it. In the case of sweet pepper fruit, with an estimated yield of 1.9 million tons/year in Europe, reports indicate that while an average time of 6 s per fruit is required for automated harvesting, the available technology has only achieved a success rate of 33% with an average picking time of 94 s per fruit [1]. For cucumber harvesting, a cycle time of 10 s was proven to be economically feasible [2]. Only in Washington State, 15–18 billion apple fruits are harvested manually every year. An estimated 3 million tons of apples is reported to have been produced in Poland in 2015 [3], out of which one-third are delicate fruits and are less resistant to bruising from mass harvester machines. Also in Florida, where the current marketable yield of sweet pepper fruits in open-field cultivation is 1.6–3.0 with potential yield of 4 lb/ft^2 in passive ventilated greenhouses [4], manual harvesting is still the only solution. Therefore, development of an automated robotic harvesting should be considered as an alternative method to address the associated labor shortage costs and timeliness.

Research and development in agricultural robotics date back to 1980s, with Japan, the Netherlands, and the USA as the pioneer countries. The first studies used simple monochrome cameras for apple detection inside the canopy [5]. Advances in the sensor technology and imaging devices have led to the employment of more sophisticated devices such as infrared [6], thermal [7] and hyperspectral cameras [8], or combination of multi-sensors [9] that are adopted with novel vision-based techniques for extracting spatial information from the images for fruit recognition, localization, and tracking. Examples of some of the recent achievements include automatic fruit recognition based on the fusion of color and 3D feature [10], multi-template matching algorithm [11], and automatic fruit recognition from multiple images [12]. Unlike the industrial case, an agriculture robot has to deal with different arrangement of plantings size and shapes, stems, branches, leaves, fruit color, texture, and different location of fruits and plants with respect to each other. Significant contributions have been made by different research groups to address these challenges; however, there is currently no report of a commercial robotic harvesting for fresh fruit market [13], mainly due to the extremely variable heterogeneous working condition and the complex and unpredicted tasks involved with each scenario. Some of the questions to be addressed in designing of a complete robotic harvesting are the simultaneous localization

Figure 1. Manual harvesting of fruits.

of fruit and environment mapping, path planning algorithms, and the number of detectable and harvestable fruits in different plant density conditions. The function of a robot can be separated into three main sections as sensing (i.e., fruit recognition), planning (i.e., hand-and-eye coordination), and acting (i.e., end-effector mechanism for fruit grasping) [14]. A common approach in fruit detection is by using a single view point, as in the case of a cucumber harvesting robot [15], or multiple viewpoints with additional sensing from one or few external vision sensors that are not located on the robot [16]. Other than the issues with frame transformation, this solution is not promising if the fruit is heavily occluded by the high density plant leaves [17]. Obviously, the final robot prototype needs to be relatively quicker for mass-harvest, with an affordable cost for greenhouse growers. Swarms of simple robots with multiple low-cost camera and grippers, or human-robot collaboration are the research topics to solve the facing challenges in robotic harvesting that current technology cannot overcome. These approaches can significantly improve the processing time of multiple fruit detection in the high-density plants, and provide ground truth results over time for machine learning algorithms based on human-operators experience. Research on agricultural robotics with a focus on automated harvesting of fruiting and vegetable are huge. See for example the works carried out on sweet pepper [1, 18–20], oil palm [21], cucumber [15, 22–24], apple [25], strawberry [26, 27], cherry fruit [6], citrus [28], and tomato [29]. Most of these works have used eye-in-hand look-and-move configuration in their visual servo control (**Figure 2**). Other researches are concentrated on the end-effector design [30], analysis of the robot performance in the dense obstacle environments using stability tests [31], motion planning algorithms [32], and orchard architecture design for optimal harvesting robot [33]. In addition, several software frameworks have been developed for agricultural robotics. An example includes the work of [34], in which a generic high-level functionality was provided for easier and faster development of agricultural robots. Some of the most recent advances in sensing for robotic harvesting include the works of [29, 35] which address the problem of detecting fruits and obstacles in dense foliage. Moreover, [20] and [25] have extensively explored the use of combined color distance, or RGB-D, data on apples and on sweet-peppers, respectively, while [36] present a study devoted to symmetry analysis in three-dimensional shapes for products detection on the plant.

Figure 2. Research and development in robotic harvesting of fruits with different manipulators and gripper mechanisms for: (A) citrus, (B, C) sweet pepper, (D, E) tomato, (F) cucumber, (G, H) strawberry, and (I–K) apple.

Improvement of robotic harvesting requires experimenting with different sensors and algorithms for fruit detection and localization, and a strategy for finding the collision-free paths to grasp the fruits with minimum control effort. Experiments with the actual hardware setup for this purpose are not always feasible due to time constraints, unavailability of equipment (i.e., sensors, cameras, and the robot manipulator), and the operation costs. In the other hand, some hardware setups may result in actuator saturation, or create unsafe situation to the operators and/or plants system. Simulation offers a reliable approach to bridge the gap between innovative ideas and the laboratory trials, and therefore can accelerate the design of a robust robotic fruit harvesting platform for efficient, cost-effective and bruise-free fruit picking. This research was motivated based on the sensing task in robotic harvesting, which requires delivering a robust pragmatic computer vision package to localize mature pepper fruits and its surrounding obstacles. The main objective was to create a completely simulated environment for improvement of plant/fruit scanning and visual servoing task through an easy testing and debugging of control algorithms with zero damage risk to the real robot and to the actual equipment. The research was carried out in two main phases: (i) the creation of the simulated workspace in the virtual robot experimentation platform (V-REP), and (ii) the development of communication and control architecture using the robot operating system (ROS) and MATLAB (The MathWorks Inc., Natick, MA, USA). The simulated workspace included an exact replica of the Fanuc LR Mate 200iD robot manipulator with six degrees of freedom (Fanuc America Corporation, Rochester Hills, MI), models of sweet pepper fruit and plant system, and different vision sensors were created in (V-REP). A simulated color camera attached to the end-effector of the robot was used as fruit localization sensor. ROS was used for exchanging data between the simulated environment and the real workspace via its publish-and-subscribe architecture. This provides a tool for validating the simulated results with those from experimenting with a real robot. V-REP and MATLAB were also interfaced to create two-way communication architecture for exchanging sensors and robot control messages. Data from the simulated manipulator and sensors in V-REP were used as inputs of a visual servo control algorithm in MATLAB. Results provided a flexible platform that saves in cost and time for experimenting with different control strategies, sensing instrumentation, and algorithms in automated harvesting of sweet pepper.

2. Overview of the simulation environment

Computer simulation of a complete robotic harvesting task requires: (i) CAD file setup including good replications of the plants-and-fruit scene and the robot manipulators, (ii) simulation environment and calculation modules for the manipulator candidates and platforms (i.e., inverse kinematics and path planning), (iii) different sensors setup, and more importantly (iv) algorithms for control tasks such as visual servoing and gripper control mechanism. The main simulation environment, scene objects, and calculation modules were built in the latest version of V-REP Pro Edu V3.4.0 for Linux 64 (available at www.coppeliarobotics.com), and ROS installed on Ubuntu 14.04.3 LTS. Some of the used terminal commands are summarized in **Table 1**.

http://dx.doi.org/10.5772/intechopen.73861

Commands	Description	Commands	Description
File commands		**System info**	
ls	Directory listing	date	Show the current date and time
ls -al	Formatted listing with hidden files	cal	Show this month's calendar
cd *dir_name*	Change directory to *dir_name*	uptime	Show current uptime
cd ~	Change to home	w	Display who is online
pwd	Show current directory	whoami	Who you are logged in as
mkdir *dir_name*	Create a directory *dir_name*	finger user	Display information about user
rm *file_name*	Delete file	uname -a	Show kernel information
rm -r *dir_name*	Delete directory *dir_name*	cat /proc./cpuinfo	CPU information
rm -f *file_name*	Force remove file	cat /proc./meminfo	Memory information
rm -rf *dir_name*	Force remove directory *dir_name*	man command	Show the manual for command
cp *file_name_1 file_name_2*	Copy *file1* to *file2*	df	Show disk usage
cp -r *dir_name*1 *dir_name*2	Copy dir1 to dir2;	du	Show directory space usage
mv *file_name_1 file_name_2*	Rename or move *file_name_1* to *file_name_2*		
Working with compressed files		**Shortcuts**	
tar xf file.tar	Extract the files from file.tar	Ctrl + Alt + T	Opens a new terminal window:
tar czf file.tar.gz files	Create a tar with gzip compression	Shift + Ctrl + T	Opens a new terminal tab:
tar xzf file.tar.gz	Extract a tar using gzip	Ctrl+C	Halts the current command
tar cjf file.tar.bz2	Create a tar with bzip2 compression	Ctrl+Z	Stops the current command,
tar xjf file.tar.bz2	Extract a tar using bzip2	Ctrl+D	Log out of current session, exit
gzip file	Compresses file and renames it to file.gz	Ctrl+W	Erases one word in the current line
gzip -d file.gz	Decompresses file.gz back to file	Ctrl+U	Erases the whole line
Install from source		Ctrl+R	Type to bring up a recent command
./configure	(For example, *./vrep.sh* will run v-rep)	!!	Repeats the last command
dpkg -i pkg.deb	Install a package (debian)	exit	Log out of current session
rpm -Uvh pkg.rpm	Install a package (rpm)		

Table 1. List of the most used Ubuntu terminal commands used for navigating in the simulation environment.

ROS Indigo was used to provide a bidirectional communication (information exchange) between simulated robot and cameras with the real world. Experimental packages for Fanuc manipulators within ROS-Industrial (available at http://wiki.ros.org/fanuc_experimental) were used for controlling the manipulator. This design allows reading information from the simulation scene (i.e., robot joints velocity, position, sensors, etc.) and publishes them across ROS network for further process. Results can be used by the simulation, and/or by the real robots and controllers. The image-based visual servo control was carried out in V-REP and MATLAB. For the sake of this chapter, we only provide a brief description of ROS and V-REP.

ROS is a collection of software frameworks for robot software development. It was originally developed in 2007 by the Stanford Artificial Intelligence Laboratory, and with the support of the Stanford AI Robot project. It provides a solution to specific set of problems encountered in the developing large-scale service robots, with philosophical goals summarized as: (i) peer-to-peer, (ii) tools-based, (iii) multi-lingual, (iv) thin, and (v) free and open-source [37]. From 2008 until 2013, development was performed primarily at Willow Garage, a robotics research institute/ incubator. During that time, researchers at more than 20 institutions collaborated with Willow Garage engineers in a federated development model. Since 2010, ROS has released several versions, including Box Turtle (March, 2010), C Turtle (August, 2010), Diamondback (March, 2011), Electric Emys (August, 2011), Fuerte Turtle (April, 2012), Groovy Galapagos (December, 2012), Hydro (September, 2013), Indigo (July, 2014), and Jade (May, 2015). The open-source ROS makes it possible to develop code and applications that can be shared and used in other robotic system with minimum effort. It also offers standard operating system features such as hardware abstraction, low-level device control, implementation of commonly used functionalities, message passing between processes, and package management. ROS Packages are files and folders that are built to create minimal collections of code for easy reuse. A ROS package usually includes the followings folders and files: *bin, msg, scripts, src, srv, CMakeLists.txt, manifest.xml* (**Figure 3**).

Fundamental concepts of the ROS are: Nodes, Messages, Topics, and Services. ROS works based on a "publish-and-subscribe" architecture where processes (called nodes) publish and/ or subscribe to specific topic on which information is exchanged in the form of messages (**Figure 3**). A Node is an executable file that uses ROS to communicate with other Nodes. A Message is ROS data type that is used when subscribing or publishing to a topic. Nodes can

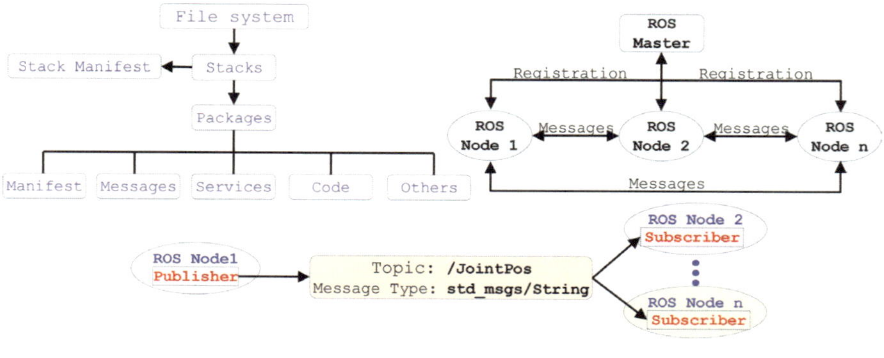

Figure 3. Diagram showing ROS file architecture and nodes communicating system.

publish messages to a Topic as well as subscribe to a Topic to receive messages. Service helps Nodes find each other. ROS nodes use a ROS client library to communicate with other nodes. Nodes can also provide or use a Service. With this architecture, each node in ROS is able to respond to input and activate other nodes, allowing participation of a sequence of nodes to complete complicated robot mission tasks. Installation details and basic configuration of ROS environment, as well as installation and configuration of packages such as V-REP/ROS bridge, and the details of the Fanuc manipulator package are not in the concept of this chapter. A more detailed discussion can be found in [38].

V-REP is like a Swiss knife in robotic simulation community. Its first public release was in March 2010, and its latest version (V3.4.0 v1) was released on April 16, 2017. It possesses various relatively independent functions, features, or more elaborate APIs, that can be enabled or disabled as desired. Compared to gazebo, V-REP is very stable and easy to set up and running. For example, the vision sensors are reasonably well simulated and if the scene is not too complex, the run times of the simulations are generally good as well. If the project requires building a custom robot in the simulator (i.e., NOVABOT or Fanuc LR Mate 200iD manipulator), the setups for links, joints, and calculation modules such as inverse kinematics necessitates some practice, however, that is the case in any robot simulation software. Another big advantage is its true cross-platform, which means it can be run in Windows or Linux. By default, the V-REP distribution for Linux should be automatically ROS enabled based on ROS

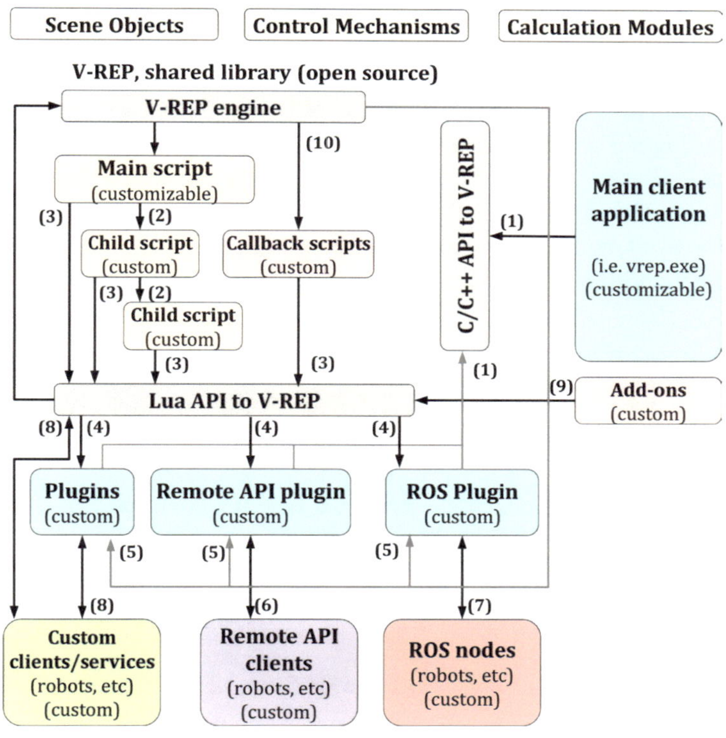

Figure 4. Schematic diagram showing the architecture and the main elements of V-REP simulator.

Indigo and Catkin. Each object/model in V-REP scene can be individually controlled via an embedded script, a plugin, a ROS node, a remote API client, or a custom solution. Controllers can be written in C/C++, Python, Java, Lua, Matlab, and Octaveor Urbi. The three main elements of V-REP simulator are scene object (i.e., joints, shape, sensors, path, etc.), calculation modules (i.e., inverse kinematics, collision detection, etc.), and control mechanism (i.e., scripts, plugin, sockets, etc.). In addition, V-REP inverse kinematics supports four different dynamic engines: The Bullet, ODE, Newton, and the Vortex Dynamics Engine. An overview of V-REP framework architecture is shown in **Figure 4**.

3. Image processing, publishing and subscription

Quantification of fruits to estimate the time required for robotic harvesting is an intensive labor task that is either ignored in high density greenhouses or is carried out manually by the use of hand pickers. We proposed a low-cost robust sweet pepper fruit recognition and tracking system using stream RGB images. Main hardware and software components of the system included a laptop computer (Lenovo Intel(R) Core(TM) i5-6200 U CPU@2.30GHz, RAM 8.00GB, 64-bit OS Windows 10), a Logitech camera (C920 HD Pro USB 1080p), supplementary halogen lamps, Adafruit Ultimate GPS breakout module 66 channel w/10 Hz (NY, USA), and Arduino Uno Microcontroller board. The image processing algorithm was implemented in MATLAB and applies median filter and image segmentation method to remove color noise from the RGB images of pepper fruits taken in the lab experiments at different angles, positions, and light conditions disturbances (varying illumination and overlapping). **Figure 5** shows: (A) original image, (B–D) red, green, and blue bands, (E) mask of only red object, (F) regions filled, (G) masked-red image, (H) extracting red component from the masked red image and applying median filter to filter out the noise, (I) convert the resulting grayscale image into a binary image and removing all pixels with a gray level value less than 3000, (J) masked image showing only red-detected object, (K) blob analysis, bounding the red objects in rectangular box and showing centroid. The image processing algorithm was validated using 78 images obtained from lab experiments and

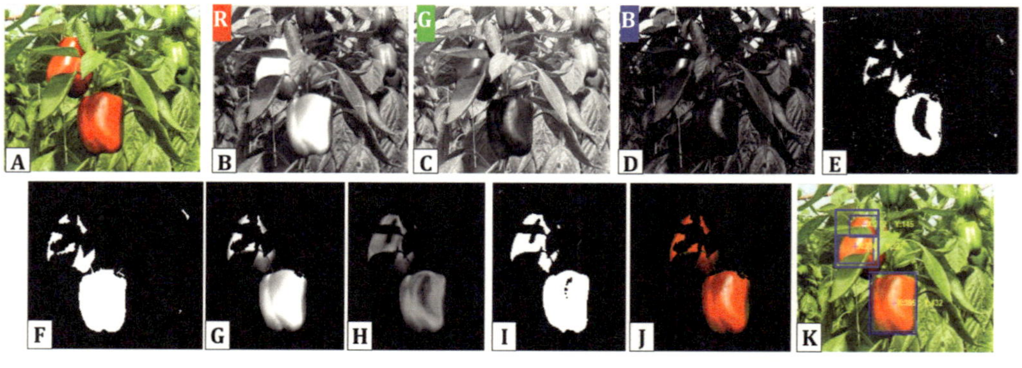

Figure 5. Demonstration of the steps in the robust image processing algorithm using edge detection with fuzzy-logic for identification and tracking of sweet pepper.

internet sources, with a recognition success rate of 94% and average recognition time of less than 2 s per image. Results of the image processing were sent from MATLAB to V-REP for simulation of visual servo control. For the actual experiment, color images of sweet pepper were acquired under natural daylight condition in different greenhouse environment in the presence of the halogen lamps. Each band of the RGB image was transferred as a 24-bit, 640 by 480 pixels matrix and was processed in real time by the custom built MATLAB application on the laptop computer. ROS was used for exchanging data between the simulated environment and the real workspace via its publish-and-subscribe architecture. Another 57 images were obtained for experimenting with different fruit and plant position scenarios. In addition, internet searched images of sweet pepper taken at different greenhouse environments were used to verify the reliability and to improve the accuracy of the algorithm. The image subscription and publishing was performed by having V-REP ROS enabled based on ROS Indigo and Catkin build. The general ROS functionality in V-REP is supported via a generic plugin *"libv_repExtRos.so"* or *"libv_repExtRos.dylib."* It should be noted that plugins are loaded when V-REP is launched, and the ROS plugin will be successfully loaded and initialized only if *"roscore"* is running at that time. The plugin is open source and can be modified as much as needed in order to support a specific feature or to extend its functionality. Three of the main ROS package folders in the V-REP, (located in *programming/ros_packages*) are the *"vrep_common,"* *"vrep_plugin,"* and *"vrep_joy"* as shown in the left side of **Figure 6**.

The first package was used to generate the services and stream messages that were needed to implement the V-REP API functions, while the second is the actual plugin that was compiled to a *".so"* file used by V-REP. The *"vrep_joy"* package enables interaction with a joystick. Having the services and stream messages in a separate package allows for other application to use them in order to communicate with V-REP via ROS in a convenient way. These packages were copied to

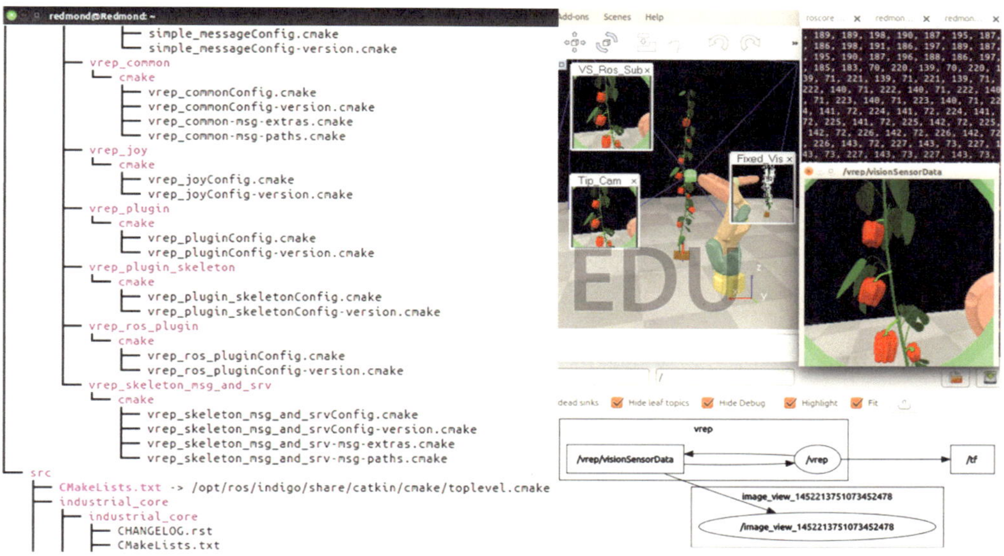

Figure 6. Image publishing and subscribing in ROS, Left image: snapshot of the main ROS package folders in the V-REP, and right image: snapshot of the simulation environment in V-REP publishing an image to a ROS node.

the *catkin_ws/src* folder. The command "*$ roscd*" was then used to check whether ROS is aware of these packages (e.g., *$ roscd vrep_plugin*). After navigating to the *catkin_ws*, the command "*$ catkin_make*" was used to build the packages and to generate the plugins. The created plugins were then copied to the V-REP installation folder to be used for image subscription and publishing. A new terminal was opened in Ubuntu for staring the ROS master using the command "*$ roscore*." Another terminal was opened and was navigated to the V-REP installation folder to launch the V-REP simulator in Ubuntu by typing the command "$. /vrep.sh." The entire procedure is summarized as these steps: (i) installing ROS Indigo on Ubuntu and setting up the workspace folder, (ii) copying "ros_packages" in V-REP into the "*catkin_ws/src*" folder, (iii) source "*setup.bash*" file, (iv) run "*roscore*" and "*. /vrep.sh*." The two available nodes, "*/rosout*" and "*/vrep*" and the three topics "/rosout," "/rosout_agg," "/vrep/info" were checked using "*$ rosnode list*" and "*$ rostopic list*" commands, respectively. In addition, the command "*$ rosservice list*" was used to advertise all the services. It should be noted that the only V-REP topic that was advertised was "*info*" publisher that started as soon as the plugin was launched. All other V-REP topics for publishing and subscribing images and sensors were individually enabled using Lua commands: "*simExtROS_enablePublisher*" and "*simExtROS_enableSubscriber*." Moreover, to visualize the vision sensor stream images and data, the "*$ rosrun image_view image_view image:=/vrep/visionSensorData*" and "*$ rostopic echo/vrep/visionSensorData*" were used, respectively. Snapshot of the simulation environment is shown in the right side of **Figure 6**.

4. Simulation scene and objects

Simulation scene in V-REP contains several elemental objects that are assembled in a tree-like hierarchy and operate in conjunction with each other to achieve an objective. In addition, V-REP has several calculation modules that can directly operate on one or several scene objects. Major scene objects and modules used in the simulation scene include (i) sensors, (ii) CAD models of the plant and robot manipulator, (iii) inverse kinematics, (iv) minimum distance calculation, (v) collision detection, (vi) path planning, and (vii) visual servo control. Other objects that were used as basic building blocks are: dummies, joints, shapes, graphs, paths, lights, and cameras (**Figure 7**). In this section, we provide description for the sensors and CAD models, and assign the next section to the calculation modules.

4.1. Sensors

V-REP supports different vision sensors (orthographic and perspective type) and proximity sensors (Ray-type, pyramid-type, cylinder-type, disk-type, and cone- or randomized ray-type

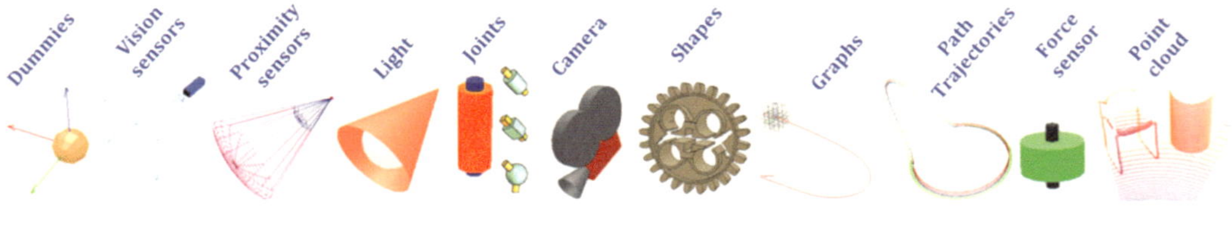

Figure 7. Major scene objects used in the simulation.

proximity sensors). It is possible to model almost any proximity sensor subtype, from ultrasonic to infrared. In addition it has built-in CAD models of several available commercial sensors such as Microsoft Kinekt, 2D and 3D laser scanners, blob detection camera, Hokuyo URG 04LX UG01, SICK S300, and TimM10 sensors. Other models can be built similarly based on combinations of different vision and proximity sensors. The V-REP model of each sensors used for this simulation is shown below its actual images in **Figure 8** which include: Fish-eye RGB Axis 212 PTZ sensor (**Figure 8A**), Infrared Proximity Sensor Long Range-Sharp GP2Y0A02YK0F (**Figure 8B**), SICK TiM310 fast laser measurement scanner (**Figure 8C**), Fast Hokuyo URG-04LX-UG01 scanning Laser Rangefinder (**Figure 8D**), and Microsoft Kinect (**Figure 8E**).

The fish-eye RGB camera was added for fruit detection, tracking, and for visual servo control with a custom set of filters that were designed for the image processing algorithm in MATLAB and V-REP. Two color cameras were also added for tracking the scene and the position of the robot end-effector with respect to the fruit and plant in order to provide a wider view of the vision sensor. The V-REP model of the Microsoft Kinect sensor includes RGB and depth vision sensors, and was used in the scene to calculate the time needed for the laser signal to hit an object and bounce back to its source, creating in this way a three-dimensional representation of the object. Five different proximity sensors with different shapes were also experimented in the simulation, including: laser ray, pyramid, cylinder, disk, and randomized ray-type. The laser-scanner rangefinder was considered in the simulation to measure distance between an observer object (i.e., the robot gripper or the end-effector camera) and a target (i.e., fruit, plant, or obstacles). Typical range finders work based on time-of-flight (TOF) and frequency phase-shift technologies. The TOF method utilizes laser by sending a pulse in a narrow beam toward the object and measuring the time taken by the pulse to be reflected off and return to the sensor. The frequency-phase shift method measures the phase of multiple frequencies on reflection together with performing simultaneous math calculations to deliver the final measure. Rangefinders are available in V-REP in the form of vision-sensors and proximity sensors. For example, the Hokuyo URG-04LX-UG01 and the 3D laser scanner range finder use a ray-type laser proximity sensor. The V-REP model for Fast-3D laser scanner uses vision sensor with the filters as illustrated in **Figure 9**. It should be noted that vision-sensors-based rangefinders have high calculation speed but lower precision, while proximity-sensors-based rangefinders have higher prevision in calculating the geometric distance with relatively lower calculation speed.

Figure 8. Major sensors used in the experiments, first row image are the actual and the second row images are the simulated sensors.

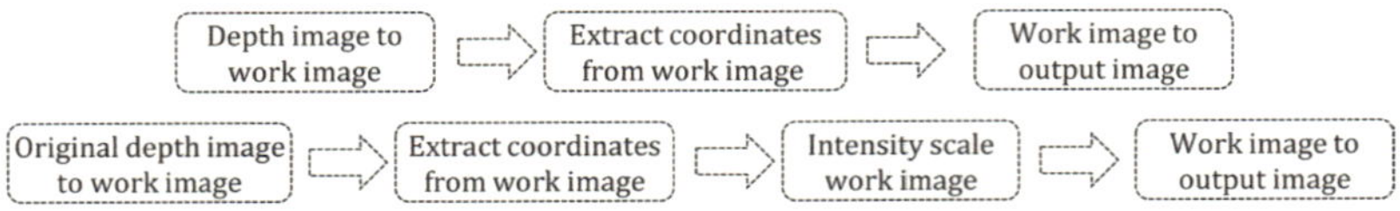

Figure 9. Filter used by each of the two vision sensors in "Fast Hokuyo URG-04LX-UG01" V-REP model (top), and by the vision sensor in the "3D laser scanner Fast" V-REP model (bottom).

4.2. CAD models

The CAD models of the sweet pepper plant, including stem system, leaves, and pepper fruits, as well as the single and multiple arms robot manipulators that were used in the simulation are shown in **Figures 10–13**. The Fanuc LR Mate 200iD robot manipulator shown in **Figure 11** is a compact six-axis robot with the approximate size and reach of a human arm. It combines best-in-class robot weight-load capacity with standard IP67 protection and outstanding FANUC quality. This makes the Fanuc LR Mate 200iD the best and most reliable mini robot for process automation in many industries. The maximum load capacity at wrist = 7 kg, repeatability = 0.02 mm, mechanical weight = 25 kg, and reach = 717 mm. The joints motion range and maximum speed are summarized in the operator manual [39]. As alternative innovative solutions, simple robots, including a platform with multiple linear actuators (**Figure 12**), and multiple SCARA robot arms (**Figure 13**) with multiple lower-cost cameras and grippers were also designed for simulation.

Figure 10. CAD models of the entire plant system: fruit, leaves, stem, calyx, and leaves.

Figure 11. Simulation scene setup with CAD models of the professional robot manipulator used in visual servo control experiment, left: Fanuc LR Mate 200iD, right: NOVABOT.

http://dx.doi.org/10.5772/intechopen.73861

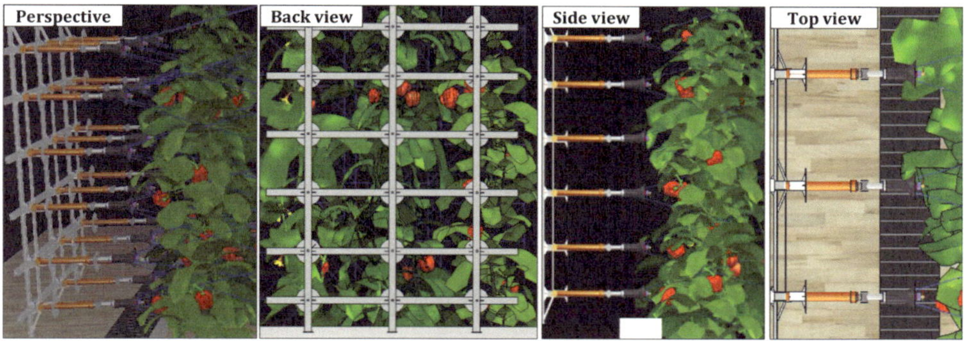

Figure 12. Simulation scene setup with CAD models of the multiple linear actuator robotic platform.

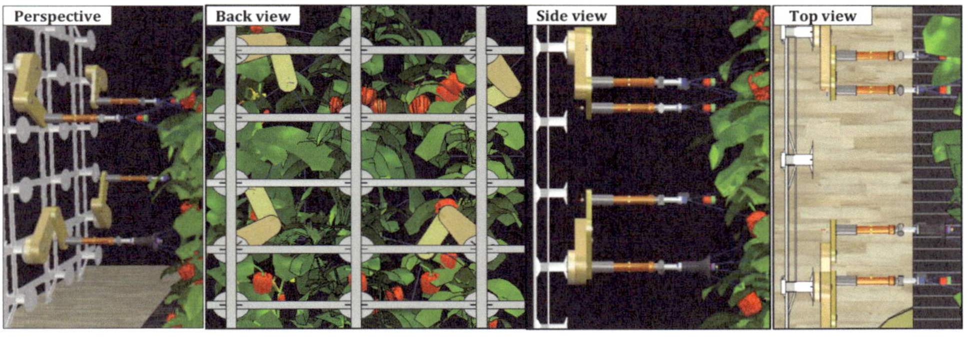

Figure 13. Simulation scene setup with CAD models of the multiple SCARA arm robotic platform.

5. Calculation modules

In order to setup the robot manipulator for different experiment, several calculation modules, including minimum distance calculation, collision detection, path planning, inverse kinematics, and different control mechanism were used in V-REP. Snapshot of the calculation modules is provided in **Figure 14**. V-REP control mechanism are divided into (i) local interfaces, including Embedded scripts, Plugins, Add-ons, and (ii) remote interfaces, including remote API clients, custom solutions, and ROS nodes, as shown in **Figure 14A**. It should be noted that different V-REP control mechanisms can be used simultaneously in one scene, or even work in conjunction with each other, which provides a multipurpose and accessible framework for the purpose of more complex robotic simulation. Scripting in V-REP is in the Lua language which is a fast scripting language designed to support procedural programming. Scripts in V-REP are the main control mechanism for a simulation. For the sake of this book chapter, we only provide brief illustration of the inverse kinematic task for the NOVABOT manipulator and the visual servo control.

5.1. Inverse kinematics

The inverse kinematic (IK) task in V-REP requires three things: (i) CAD data of the manipulator links (ii) joints, (iii) kinematic chain, (iv) tip and target dummies, and (iv) IK task. The CAD

Figure 14. Demonstration of (A) five different control mechanisms in V-REP, (B) inverse kinematics chain, (C, D) minimum distance calculation from tip vision sensor and plant/fruit model, (E) collision detection between robot links and plant model, and (F) path planning for moving a harvested fruit to the bin.

file was imported to the scene from [Menu bar --> File --> Import --> Mesh]. It should be noted that depending on how the original CAD data was generated in the original CAD software, the imported mesh file could be at a different scale, different location, or even subdivided into several shapes. The assigned color of imported shapes is random. V-REP also provides basic tools and options for creating model of a new robot if the CAD file is not available from external sources. Upon importing the CAD file, a single simple shape is located in the middle of the scene and appears in the scene hierarchy on the left hand side of the main window. For the IK task, the single CAD shape was divided by selecting [Menu bar --> Edit --> Grouping/Merging --> Divide selected shapes]. This divided the original shape into several sub-shapes that were grouped manually for a same rigid entity using [Menu bar --> Edit --> Grouping/Merging --> Group selected shapes]. For example, all shapes that were related to the robot base were grouped together and renamed as *robot_base* in the scene hierarchy. It is usually easier to change the color of each shape for a better visual appearance and for selecting the shapes that belong to one group. In the case, when all shapes that were meant to be grouped shared the same visual attributes, they were merged together instead using [Menu bar --> Edit --> Grouping/Merging --> Merge selected shapes]. After the shapes were grouped in a compound shape, the robot joints that logically belong to a shape (robot link) were added into the scene using [Menu bar --> Add --> Joint --> Revolute] with their correct position and orientation specified. All joints were then set to the IK mode and were placed at the correct position. In case, when the exact joint positions were not known, they were extracted manually based on the position of the relevant shapes. It is often helpful to refer to the robot design manual for a better understanding of links and joints functionality for building the kinematic chain, going from tip to base. The IK task requires specification of the kinematic chain described with a "tip" dummy and a "base" object, and a "target" dummy that the "tip" dummy will be constrained to follow as shown in **Figure 15**. After all elements for the definition of the IK task were ready, the "target dummy" was selected as the linked dummy to the "tip dummy," and the IK task was registered as an IK group with proper selection of calculation method (DLS or pseudo inverse), damping, and constraints (x, y, z, alpha-beta, gamma).

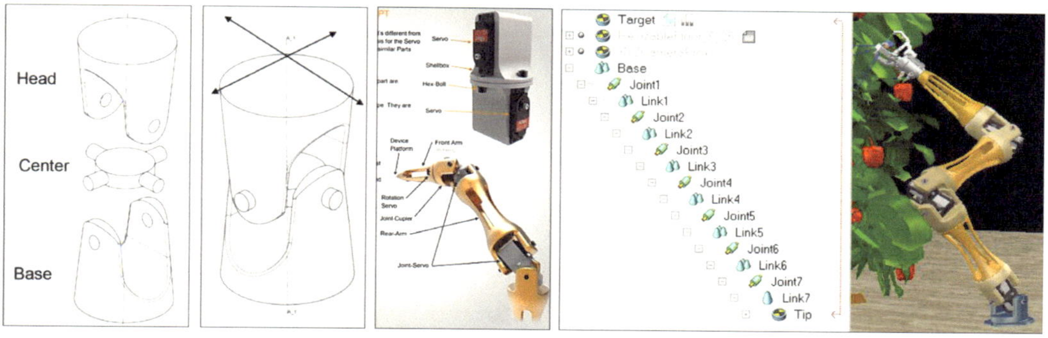

Figure 15. Demonstration of the joint functionality and the inverse kinematics chain for the NOVABOT manipulator.

5.2. Visual servo control algorithm

A robot can be controlled in V-REP simulation through several ways such as child script, writing plugins, ROS nodes, external client applications that relies on the remote API, or writing an external application that communicates with V-REP plugin or script via pipes, sockets, or serial port. V-REP supports seven supported languages: C/C++, Python, Java, Matlab, Octave, Lua, and Urbi. In this research, we used MATLAB as the remote API because it provides a very convenient and easy way to write, modify and run image based visual servoing control codes. This also allows controlling a simulation or a model (e.g., a virtual robot) with the exact same code as the one that runs the real robot. The remote API functionality relies on the remote API plugin (on the server side), and the remote API code on the client side. Both programs/projects are open source (i.e., can be easily extended or translated for support of other languages) and can be found in the 'programming' directory of V-REP's installation. Visual servo control scheme with eye-in-hand configuration, as shown in **Figure 16**, was implemented in MATLAB

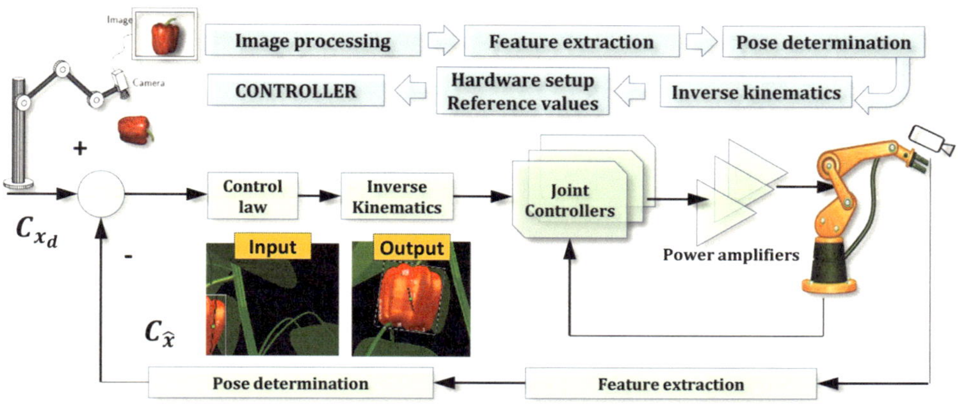

Figure 16. Visual servo control scheme with eye in hand configuration based on image moment method used with the Fanuc LR Mate 200iD and the NOVABOT manipulators.

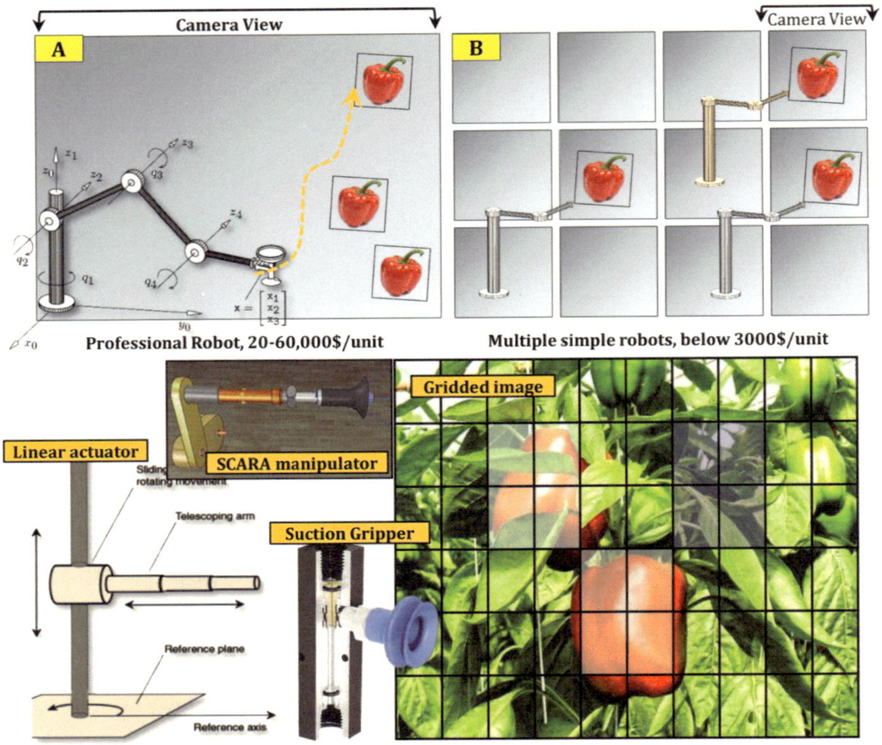

Figure 17. Schematic diagram of an innovative approach for robotic harvesting based on multiple low-cost manipulators (e.g., multiple linear actuators or SCARA arms).

based on image moment method. For the case of the multiple linear actuators and the SCARA arms, we divided the camera view into multiple camera views to enhance the accuracy of the fruit detection algorithm and also to accelerate the image processing time (**Figure 17**). Details of the visual servo control algorithm are considered intellectual property of authors' research group and are beyond the content of this chapter.

6. Results and discussions

Results provided a simulated environment for improvement of plant/fruit scanning and visual servoing task through easy testing and debugging of control algorithms with zero damage risk to the real robot and to the actual equipment. It also contributed to experimenting new ideas in robotic harvesting of sweet pepper, as well as testing different sensing instrumentation and control strategies on the currently used manipulators. Three groups of experiments, with separated V-REP scenes were designed for investigating different algorithms, robot manipulator, and sensors setup. They are summarized as experimenting with: (i) fruit detection and tracking algorithm using different camera views (**Figures 18** and **19**), (ii) manual and automated plant/fruit scanning in x-y, x-z, and y-z plane, and x-y-z space (**Figures 20** and **21**),

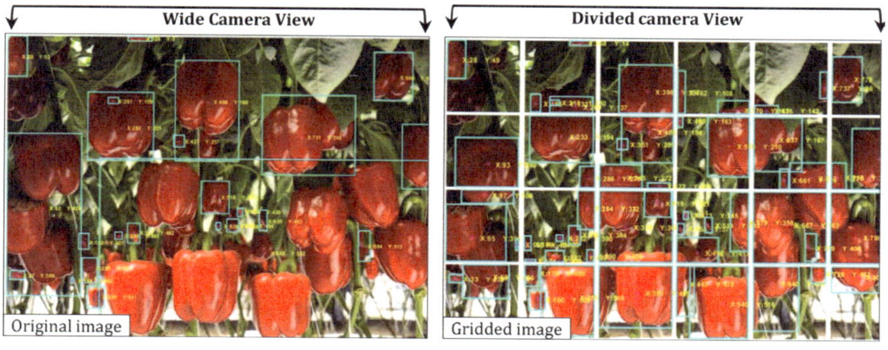

Figure 18. Results of the image processing algorithm for fruit localization using wide camera view (left) and divided camera view (right).

(iii) fruit/plant scanning using Kinect, Hokuyo, fast 3D Laser, proximity 3D Laser scanner, and proximity Hokuyo URG04LXUG01 sensors (**Figure 22**), and (iv) visual servoing control law on single (**Figure 23**) and multiple (**Figure 24**) robot manipulator. Depending on the objectives of each scenario, sensors were placed in fixed spots, or on a moving link of the robot such as the end-effector. For example, the RGB vision sensor for fruit detection and tracking was used as eye-in-hand configuration with end-point closed-loop control. For the manual fruit/plant scan experiment with RGB sensors, the robot joints were controlled via sliders or by directly entering the desired joint angles in each label box as shown in **Figure 20**. This enabled sensing from the gripper from multiple viewpoints. In order to provide an interface with real workspace, two 2-axis analog Joysticks with push button were then used with Arduino Uno microcontroller to manually control angular positions for the joints. The automated fruit/plant

Figure 19. Result of the image processing algorithm for quantification and tracking of sweet pepper in different fruit-and-plant scenario.

Figure 20. Two dimensional scanning experiment (x-y, x-z, and y-z planes) for finding the maximum fruit visibility. Camera was set to move at 30 degrees increments around the fruit and plant model.

scan experiments with RGB sensor were also carried out in different x, y, and z direction. The objective from this experiment was to simulate various camera pose and views for the best fruit attack and harvest. For Scanning in x-y plane, a 360° scan configuration of the fruit in the horizontal x-y plane is shown in **Figure 20**, with 30° increment snapshots of the simulated fruit. A similar scanning has been employed by [40]. For scanning in x-y-z space, two scan configurations in x-y-z space were used with snapshots of the resulting camera view shown in **Figure 21**. In this setup, the RGB sensor mounted on the robot tip is moved on the horizontal plane x-y to find the best view of the fruit. Moreover, the manipulator is "twisted" to provide different viewpoints for the end-effector camera.

The "3D Laser Scanner Fast" sensor model in V-REP is based on vision-sensor with a perspective angle equal to 45°, a resolution of 512 by 512 and minimum and maximum 0.05 and 5 m distance of operation. Snapshot of the experiment with this sensor is shown in **Figure 22**. The "Fast Hokuyo URG-04LX-UG01" model in V-REP also works in perspective mode with an operability angle equal to 120°, and a resolution that was set at 512 by 1 which means it scans along a line shown on the floating view. It has a minimum and maximum distance of operability, respectively, equal to 0.04 and 5. The image processing in this case is similar to the 3D laser sensor except that the intensity map scale component is omitted. This sensor in fact does not come with any floating view by default. Two floating views were added for the

Figure 21. Three dimensional scanning experiments (x-y-z space) for finding the maximum fruit visibility. Camera was set to rotate around the fruit and plant until it finds the best angle of attack.

http://dx.doi.org/10.5772/intechopen.73861

Figure 22. Simulation of scanning experiment with Laser scanners and depth sensors.

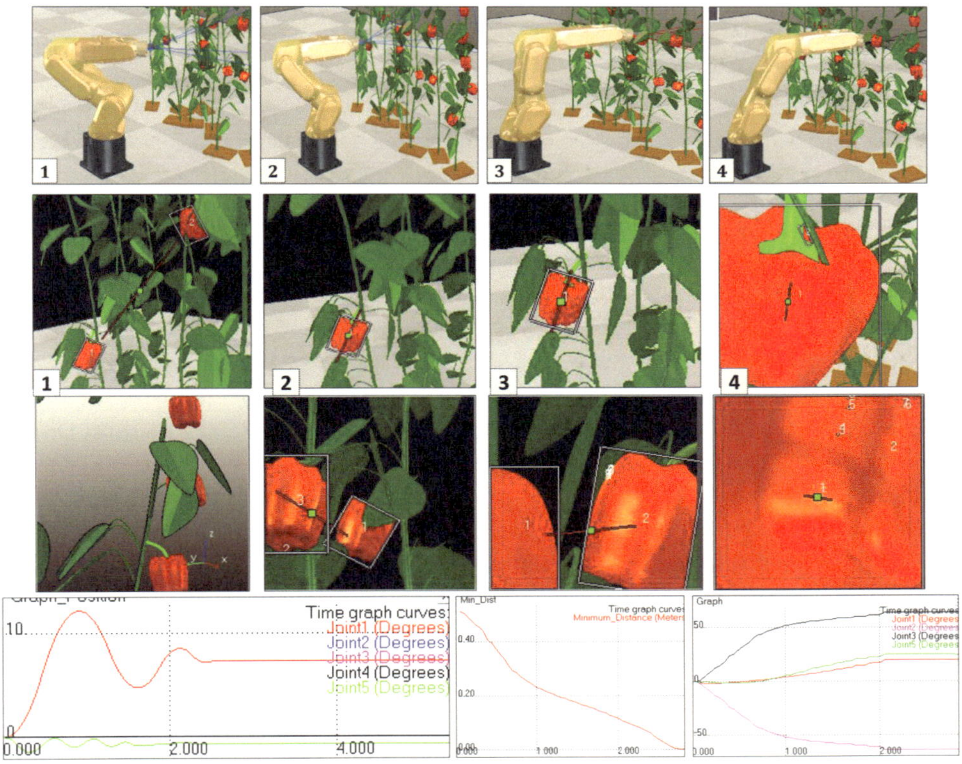

Figure 23. Simulation of visual servo control experiment with the eye-in-hand configuration and PID Control law on joint angles with feedbacks from image moments. Stability was achieved in 2.5 s without overshoot and oscillations.

Figure 24. Simulation of robotic harvesting with arrays of linear actuators.

two vision sensors of the "Fast Hokuyo URG-04LX-UG01" model. The black line inside the floating view of each sensor indicates presence of object (i.e., fruit, leaf, robot, and plant). First row images in **Figure 22** are snapshot of the scene with Kinect depth sensor in action for fruit/plant scan, and the bottom row images are, respectively, from left to right are: snapshot of the scene with vision sensors, showing the "3D Laser scanner Fast," the "Fast Hokuyo URG-04LX-UG01," snapshot of the scene with proximity sensors showing the "3D-Laser scanner," and the "Hokuyo URG-04LX-UG01" scanning scene.

For the visual servo control task, a robot end-mounted camera was used to position the robot arm in a plane orthogonal to the axis, such that the fruit to be harvested is centered in the camera's field of view. The system had no trajectory generator, instead a feedback loop closed visually, was used to control the robots arm position. The sensor and robot was programmed for visual servoing task in such a way that the end-effector tracks the largest detected fruit until maximum possible view of that fruit is provided. Two different control approaches was designed and tested, one based on joint velocity control and the other based on joint position control. In both design a PID control law was implemented to minimize the offset error between image position of a detected fruit and center of the camera frame. Results showed that the robot could adjust itself in such a way that its tip RGB sensor shows maximum possible view of the largest detected fruit and become stable in a short time. It should be noted that both control algorithms were designed and tuned based on the experiments and statistical results from fruit/plant scan. Video demonstration of the entire experiments can be accessed from the links listed in **Table 2**.

As the final discussion, we would like to highlight that agricultural robotic is a part of the big picture in the future production of vegetable and crops, i.e., growing plants in space.

Simulation experiment	Video demo link
Simulation of NOVABOT innovative manipulator	https://youtu.be/R38IoVcOVt0
Simulation of multiple SCARA arms	https://youtu.be/TLLW3S-55ls
Simulation of multiple linear actuators	https://youtu.be/iFu7FAxLvmg
Robotic Harvesting with Fanuc LR Mate 200iD	https://youtu.be/BwRBUeB812s
Simulation of Environment mapping and scanning	https://youtu.be/XD3J7b0cDGM
Detailed demonstration of fruit and plant scan	https://youtu.be/6EOy1NesvQg
Detailed demonstration of visual servo control	https://youtu.be/VupoirQOL0Y
Testing Visual Servo Control Algorithm	https://youtu.be/VupoirQOL0Y
Environment Setup: Ubuntu, V-REP, ROS	https://youtu.be/tKagjNQ9FW4
Real-time, robust and rapid red-pepper fruit detection	https://youtu.be/rFV6Y5ivLF8

Table 2. Links to the video demonstrations.

An example includes space research for development of Mars greenhouses to produce vegetables during a mission to Mars or at Antarctica. The trend in food production is toward urban farming techniques, compact Agri-cubes, and automated systems with minimum human interface. The idea is that even people with limited experience/knowledge in vegetable cultivation can operate these units. While this integration might seem too ambitious, it can serve as a prophetic awareness for a perceptive outlook in the farming system. For example, the conventional arrangements of citrus groves, orchards, and the trees shapes in Florida had to be reconsidered for the mechanical harvesting machines to operate successfully. It is likely that the greenhouse planting systems for sweet pepper will also be re-shaped to match with a customized robotic platform. Two of the key challenges to be solved during the design of robotic harvesting framework are addressed by [40] as (i) detection of a target location of the fruit, and (ii) moving the end-effector toward that location with precision for harvesting task. We argue that these two challenges have not been stated accurately. First of all, it is not always necessary to detect the target location of the fruit, especially in the case of a mass harvesting platform with shaking grippers. Second, moving the end-effector toward the target fruit is not a scientifically sound statement, e.g., considering the strategy case in which the plant system is moved toward a fixed end-effector. To avoid this divergence, the task should have been stated as minimizing the error between the location of the end-effector and the target fruit. In fact, a promising solution to the robotic harvesting is not through a single robot manipulator. We provided a quick review of the previous works, and used simulation approach to reveal that single arm robots for harvesting are still far beyond realization, and have failed mainly due to the sensing and moving action in high vegetation density. In this approach, even if the fruit localization is accurate, and the robot control calculates an optimum trajectory to reach the fruit without receiving additional sensing feedback from the camera, the moment it enters into the dense plant canopy it disrupts the exact location of the target fruit.

7. Conclusion

Research and development for the use of robotics in agriculture that can work successively have grown significantly in the past decade; however, to this date, a commercial robotic harvesting is still unavailable for fresh fruit market. With the decrease of greenhouse workforce and the increase of production cost, research areas on robotic harvesting have received more and more attention in recent years. For the success of robotic harvesting, the identification of mature fruit and obstacle is the priority task. This chapter reported on a simulation and control platform for designing, testing, and calibration of visual servoing tasks in robotic harvesting of sweet-pepper. Creation of a virtual environment was carried out as a response to the improvement of fruit detection rate. We provided a documented guideline for a reliable, cheap and safe experiment platform with a faster approach for development, testing, and validating control strategies and algorithms to be used with different robot candidates and gripper mechanism in automated harvesting of fruiting vegetables. Results of the image processing confirmed that the presented approach can quantify and track mature red sweet pepper fruits from its surrounding obstacles in the real-time. It can be concluded that development of an affordable and efficient harvesting robot requires collaboration in areas of horticultural engineering, machine vision, sensing, robotics, control, intelligent systems, software architecture, system integration, and greenhouse crop management. In addition, practicing other cultivation systems in the greenhouse, such as single row, might be necessary for overcoming the problems of fruit visibility and accessibility. It can also be concluded that human-robot collaboration might be necessary to solve the challenges in robotic harvesting that cannot yet be automated. In a collaborative harvesting with human-robot interface, any fruit that is missed by the robot vision is identified by the human on the touch screen, or the entire robot actions are controlled manually in a virtual environment. Nevertheless, robotic harvesting must be economically viable which means it must sense fast, calculate fast, and move fast to pick a large number of fruits every hour that are bruise free.

Author details

Redmond R. Shamshiri[1,3], Ibrahim A. Hameed[1]*, Manoj Karkee[2] and Cornelia Weltzien[3]

*Address all correspondence to: ibib@ntnu.no

1 Department of ICT and Natural Sciences, Faculty of Information Technology and Electrical Engineering, NTNU, Ålesund, Norway

2 Department of Biological Systems Engineering, Center for Precision and Automated Agricultural Systems, IAREC, Washington State University, Prosser, WA, United States

3 Leibniz Institute for Agricultural Engineering and Bioeconomy, Potsdam-Bornim, Germany

http://dx.doi.org/10.5772/intechopen.73861

References

[1] Hemming J, Bac W, van Tuijl B, Barth R, Bontsema J, Pekkeriet E, van Henten E. A robot for harvesting sweet-pepper in greenhouses. In: Proceedings of the International Conference on Agricultural Engineering; 2014. pp. 6-10

[2] Van Henten EJ, Van't Slot DA, Hol CWJ, Van Willigenburg LG. Optimal manipulator design for a cucumber harvesting robot. Computers and Electronics in Agriculture. 2009; **65**(2):247-257

[3] Młotek M, Kuta Ł, Stopa R, Komarnicki P. The effect of manual harvesting of fruit on the health of workers and the quality of the obtained produce. Procedia Manufacturing. 2015;**3**:1712-1719

[4] Vallad GE, Smith HA, Dittmar PJ, Freeman JH. Vegetable production handbook of Florida. Gainesville, FL, USA: University of Florida, IFAS Extension; 2017. Available at: http://edis.ifas.ufl.edu/pdffiles/CV/CV29200.pdf [Last access: February 10th, 2018]

[5] Peilin L, Lee S, Hsu H-Y. Review on fruit harvesting method for potential use of automatic fruit harvesting systems. Procedia Engineering. 2011;**23**:351-366

[6] Tanigaki K, Fujiura T, Akase A, Imagawa J. Cherry-harvesting robot. Computers and Electronics in Agriculture. 2008;**63**(1):65-72

[7] Bulanon DM, Burks TF, Alchanatis V. Study on temporal variation in citrus canopy using thermal imaging for citrus fruit detection. Biosystems Engineering. 2008;**101**(2):161-171

[8] Okamoto H, Lee WS. Green citrus detection using hyperspectral imaging. Computers and Electronics in Agriculture. 2009;**66**(2):201-208

[9] Bulanon DM, Burks TF, Alchanatis V. Image fusion of visible and thermal images for fruit detection. Biosystems Engineering. 2009;**103**(1):12-22

[10] Tao Y, Zhou J. Automatic apple recognition based on the fusion of color and 3D feature for robotic fruit picking. Computers and Electronics in Agriculture. 2017;**142**(Part A):388-396

[11] Bao G, Cai S, Qi L, Xun Y, Zhang L, Yang Q. Multi-template matching algorithm for cucumber recognition in natural environment. Computers and Electronics in Agriculture. 2016;**127**(Supplement C):754-762

[12] Song Y, Glasbey CA, Horgan GW, Polder G, Dieleman JA, van der Heijden GWAM. Automatic fruit recognition and counting from multiple images. Biosystems Engineering. 2014;**118**(Supplement C):203-215

[13] Bac Y, Henten CW, Hemming EJ, Edan J. Harvesting robots for high-value crops: State-of-the-art review and challenges ahead. Journal of Field Robotics. 2014;**31**(6):888-911

[14] Murphy R. Introduction to AI robotics. Cambridge, MA, USA: MIT Press; 2000

[15] Van Henten J, Van Tuijl EJ, Hoogakker BAJ, Van Der Weerd GJ, Hemming MJ, Kornet J, Bontsema JG. An autonomous robot for de-leafing cucumber plants grown in a high-wire cultivation system. Biosystems Engineering. 2006;**94**(3):317-323

[16] Hemming J, Ruizendaal J, Illem Hofstee JW, van Henten EJ. Fruit detectability analysis for different camera positions in sweet-pepper. Sensors (Basel). 2014;**14**(4):6032-6044

[17] Bac CW. Improving obstacle awareness for robotic harvesting of sweet-pepper. Wageningen, The Netherlands: Wageningen University; 2015

[18] Hemming J, Bontsema J, Bac W, Edan Y, van Tuijl B, Barth R, Pekkeriet E. Final Report on Sweet-Pepper Harvesting Robot; December; 2014. p. 22

[19] Bac CW, Hemming J, Van Henten EJ. Robust pixel-based classification of obstacles for robotic harvesting of sweet-pepper. Computers and Electronics in Agriculture. 2013;**96**: 148-162

[20] Vitzrabin E, Edan Y. Adaptive thresholding with fusion using a RGBD sensor for red sweet-pepper detection. Biosystems Engineering. 2016;**146**:45-56

[21] Shamshiri R, Ishak W, Ismail W. Nonlinear tracking control of a two link oil palm harvesting robot manipulator. 2012;**5**(2):1-11

[22] Van Henten EJ, Hemming J, van Tuijl BAJ, Kornet JG, Meuleman J, Bontsema J, van Os EA. An autonomous robot for harvesting cucumbers in greenhouses. Journal of Autonomous Robots. 2002;**13**:241-258

[23] Tang Y, Zhang X, Liu T, Xiao L, Chen D. A new robot system for harvesting cucumber. In: American Society of Agricultural and Biological Engineers Annual International Meeting; 2009. pp. 3873-3885

[24] Van Henten EA, Van Tuijl EJ, Hemming BAJ, Kornet J, Bontsema JG, Van Os J. Field test of an autonomous cucumber picking robot. Biosystems Engineering. 2003;**86**(3):305-313

[25] Thanh Nguyen W, Vandevoorde T, Wouters K, Kayacan N, De Baerdemaeker E, Saeys JG. Detection of red and bicoloured apples on tree with an RGB-D camera. Biosystems Engineering. 2016;**146**:33-44

[26] Han HH, Kil-Su S-CK, Lee YB, Kim SC, Im DH, Choi HK. Strawberry harvesting robot for bench-type cultivation. Biosystems Engineering. 2012;**37**(1):65-74

[27] Hayashi S, Shigematsu K, Yamamoto S, Kobayashi K, Kohno Y, Kamata J, Kurita M. Evaluation of a strawberry-harvesting robot in a field test. Biosystems Engineering. 2010;**105**(2):160-171

[28] Mehta SS, Burks TF. Vision-based control of robotic manipulator for citrus harvesting. Computers and Electronics in Agriculture. 2014;**102**:146-158

[29] Senthilnath J, Dokania A, Kandukuri M, Ramesh KN, Anand G, Omkar SN. Detection of tomatoes using spectral-spatial methods in remotely sensed RGB images captured by UAV. Biosystems Engineering. 2016;**146**:16-32

[30] De-An Z, Jidong L, Wei J, Ying Z, Yu C. Design and control of an apple harvesting robot. Biosystems Engineering. 2011;**110**(2):112-122

[31] Li Z, Li P, Yang H, Wang Y. Stability tests of two-finger tomato grasping for harvesting robots. Biosystems Engineering. 2013;**116**(2):163-170

[32] Bac CW, Roorda T, Reshef R, Berman S, Hemming J, van Henten EJ. Analysis of a motion planning problem for sweet-pepper harvesting in a dense obstacle environment. Biosystems Engineering. 2016;**146**:85-97

[33] Bloch V, Degani A, Bechar A. A methodology of orchard architecture design for an optimal harvesting robot. Biosystems Engineering. 2018;**166**:126-137

[34] Hellström O, Ringdahl T. A software framework for agricultural and forestry robots, Industrial Robot. Industrial Robot: An International Journal. 2013;**40**(1):20-26

[35] Amatya S, Karkee M, Gongal A, Zhang Q, Whiting MD. Detection of cherry tree branches with full foliage in planar architecture for automated sweet-cherry harvesting. Biosystems Engineering. 2015;**146**:3-15

[36] Barnea E, Mairon R, Ben-Shahar O. Colour-agnostic shape-based 3D fruit detection for crop harvesting robots. Biosystems Engineering. 2016;**146**:57-70

[37] Quigley M, Conley K, Gerkey B, Faust J, Foote T, Leibs J, Berger E, Wheeler R, Mg A. ROS: an open-source Robot Operating System. ICRA. 2009;**3**:5

[38] Bouchier P. Embedded ROS [ROS Topics]. IEEE Robotics and Automation Magazine. 2013;**20**(2):17-19

[39] FANUC Robot LR Mate 200ic mechanical unit operator's manual. MAROTLR2006071E REV. G, B-82584EN/06. Rochester Hills, Michigan: FANUC Robotics America, Inc.; 48309-3253. Available at: http://www.msamc.org/aimss/documentation/pdf/manuals/lr_mate_manuals/LR%20Mate%20200iC%20Operators%20Manual.pdf [Last access: February 10th, 20]

[40] Barth R, Hemming J, van Henten EJ. Design of an eye-in-hand sensing and servo control framework for harvesting robotics in dense vegetation. Biosystems Engineering. 2016;**146**:71-84

Chapter 6

Ubiquitous Environment Control System: An Internet-of-Things–Based Decentralized Autonomous Measurement and Control System for a Greenhouse Environment

Takehiko Hoshi, Ken-Ichiro Yasuba,
Hideto Kurosaki and Takashi Okayasu

Additional information is available at the end of the chapter

http://dx.doi.org/10.5772/intechopen.71661

Abstract

A low-cost and flexible system for environmental measurement and control in greenhouses based on decentralized autonomous technics, Ubiquitous Environment Control System (UECS), was proposed in 2004. The UECS is composed of autonomous nodes as the minimum units of measurement and control. The nodes can connect with each other through Ethernet or Wi-Fi and can communicate information regardless of manufacturer or model. To realize automation and efficiency of protected horticultural production, two consortia for UECS development and extension were established. During the last 10 years, the UECS has been used to apply environment control in large-scale greenhouses and plant factories. The stability and utility of the UECS have been demonstrated and verified in these practical cultivations. Current research and development are being carried out to install information and communication technology (ICT) systems to improve productivity in existing small- to medium-scale greenhouses in Japan. The flexibility and concept of the UECS have been very effective to enable sophisticated environmental control technology to be applied to small- and medium-scale greenhouses. In this chapter, self-fabricated UECS, the renewal of old environmental control systems using the UECS, and Sub-GHz radio band use for communicating UECS nodes among distributed greenhouses are discussed.

Keywords: protected horticulture, smart agriculture, open-source hardware, restructuring greenhouses, managing distributed greenhouses

1. Introduction

The greenhouse, which is a solar energy utilization facility, is covered with a thin transparent material, and, therefore, the greenhouse environment is greatly affected by the outside climate and solar irradiation. To control the greenhouse environment to ensure suitability for crop production, it is necessary to frequently operate facilities such as windows, heaters, and curtains. Research and development of environmental control systems intensified after low-cost computers such as minicomputers and microcomputers were invented [1, 2]. Since computerized environmental control systems were still too expensive for greenhouse crop production until the 1990s, their use and development potential was limited to large-scale and well-equipped greenhouses in the USA and European. In contrast, Asian greenhouses were small-scale and ill-equipped, so installation of environmental control systems hardly progressed after this time period.

According to Moore's Law, cost-performance of computers was sharply improved, and information and communication technologies (ICT) were also improved simultaneously. As a result, autonomous distributed computing technologies such as ubiquitous computing [3, 4], which use many networked computers, began to be introduced into various fields in the twenty-first century. In 2004, a decentralized autonomous system for environmental control in greenhouses was proposed in Japan [5]. In this system, the measurement and control elements of the greenhouse are divided into units, which are termed "nodes," and a low-cost computer system is allocated to each node. Because the computers in each node are networked with each other and measure and control the environment everywhere (ubique) in the greenhouse, the system was termed "Ubiquitous Environment Control System" (UECS). **Figure 1** shows an example of the latest UECS-applied greenhouse.

Using a flexible node-network formation, the UECS is able to measure and control the environment in a range of facilities, from small greenhouses requiring only ventilating windows control to large-scale crop production facilities, such as plant factories, that require complex

Figure 1. Greenhouse strawberry production using a hanging hydroponic bed in Okayama prefecture, Japan. Temperature, humidity, light intensity, carbon dioxide gas concentration, and nutrient solution supply are controlled precisely by the UECS.

http://dx.doi.org/10.5772/intechopen.71661

control. In addition, since the communication protocol of the UECS network is open, interconnection and interoperability are possible even for products from different manufacturers [6]. In this chapter, an overview and present research on the UECS are given in detail.

2. Overviews of UECS

2.1. Minimum unit—node

Conventional environment control systems are controlled by a single computer and concentrate on electric signal lines from all sensors (e.g. air temperature and humidity) and actuators (e.g. heater and roof ventilator) in the greenhouse and intensively perform measurements. It is an advantage that integrated control coordination of the sensors and actuators is easy to realize in such converged systems. However, some disadvantages are pointed as follows: (1) complicated control program, (2) no extensibility, (3) high risk of whole system failure due to breakdown of only the central computer. In addition, since the specifications of electrical signal lines are often not unified, the models and manufacturers of connectable instruments to the environmental control system are limited.

In contrast, a decentralized autonomous control system for the greenhouse environment, UECS, is composed of nodes that are the minimum units. The node shown in **Figure 2** has an embedded computer in one or several sensors and actuators. The computer has a communication port for networking with other nodes, and a measurement and control program for sensors and actuators.

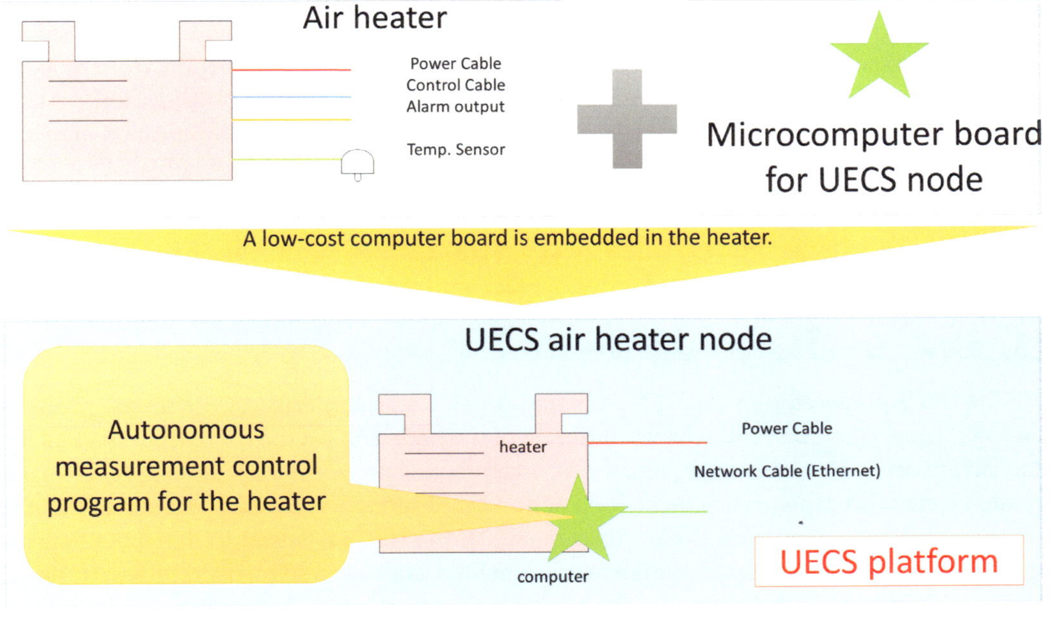

Figure 2. Schematic diagram of a node, which is the minimum unit of UECS components.

Taking the air heating node as an example, when the node does not receive a communication message (named UECS-CCM here), the node autonomously controls the air heater using the inbuilt air temperature sensor and the control set point. At the same time, the message on the air heater operation is also transmitted to the other nodes. If the remote control message is received from another node, the operation of the air heater is determined by the content of message. If data reception is stopped for a specific period, the node returns to autonomous control. If higher priority air temperature data are received, these data are used instead of the data from the inbuilt air temperature sensor. Even if one of the sensors fails, the control of the air heater is not stopped.

A measurement and control program is installed in the embedded computer for the sensors and actuators that belong to the node. As a result, the program is simplified, its development is facilitated, and even with low performance, computer boards are available to embed. The communication protocol is standardized as described in the following sections. All the UECS nodes can connect to the network and can communicate information regardless of manufacturer or model. The UECS can be configured with a free combination of the nodes, and it has high expandability. Because of the autonomous function of the nodes, even if one node fails, the risk of stopping the entire system is relatively small.

2.2. Communication protocol and user interface

The nodes of the UECS use Ethernet (IEEE 802.3) or Wi-Fi (IEEE 802.11) to compose the communication network, and they mainly use the broadcast packets of User Datagram Protocol (UDP) to exchange messages for environmental measurement and control. Our pilot study that tested a decentralized control system using Internet protocol (IP) in a greenhouse environment showed that the major risk factors were delays and stoppages of the system due to hanging up the virtual circuit of transmission control protocol (TCP). For this reason, UDP, a connectionless communication, was employed. To eliminate the complicated installation work of associating among the nodes for data exchange, or the network server for transmission control, neither unicast nor multicast packet transmission was employed. The communication message protocol for UECS, which has been named "Common Corresponding Message" (UECS-CCM), is managed and operated by UECS consortium [7].

Figure 3 shows an example of the UECS-CCM used to exchange measurement and control data. A message written in XML is added to the specific attributes for delivering the broadcast message in the DATA tag. Timings of message transmission according to kinds of information are classified into two classes, they are periodical time intervals and occasions on demands.

All the UECS nodes employ an HTTP server to provide a user interface. Greenhouse growers and managers can monitor the latest condition of the node and set the control parameters for the facility using a Web browser interface by accessing the private IP address assigned to each node. Electrical components such as switches and indicators occupy a large share of the initial cost of measurement and control instruments. If the UECS is introduced in greenhouses, growers can manage the greenhouse environmental control system using a smartphone, tablet, or a portable game console, which have higher usability as user interfaces instead of expensive electrical components (**Figure 4**).

http://dx.doi.org/10.5772/intechopen.71661


```xml
<?xml version="1.0"?>
<UECS ver="1.00-E10">
<DATA type="inAirTemp" room="1" region="1" order="1"
priority="15">23.5</DATA>
<IP>192.168.1.64</IP>
</UECS>
```


Figure 3. An example of the UECS-CCM. This message sent from the IP address of 192.168.1.64 was issued from the first node in the first compartment of the first greenhouse with the 15th priority, and the node reported that the inside air temperature was 23.5°C.

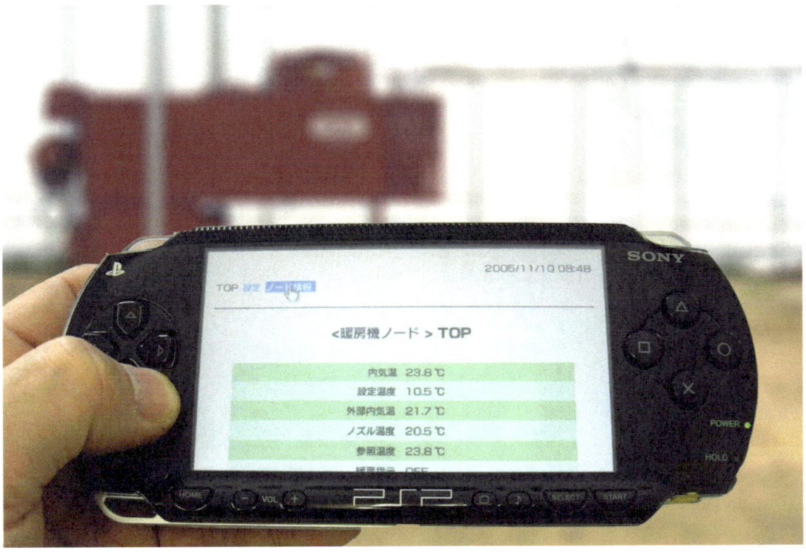

Figure 4. Managing the node for an oil air heater using a portable game console in a greenhouse using an UECS.

2.3. Case study on the introduction of an UECS in a greenhouse used for tomato production

As a case study of the early stage of UECS development, UECS was introduced for the environmental control of a greenhouse (floor area: 1782 m^2) in 2007 used for tomato production in Tokushima prefecture, Japan [6]. Overall, 16 sets of nodes from different manufacturers were connected through a LAN (**Figure 5**). The environmental control algorithm working autonomously in each node has a relatively simple function. To perform complex environmental control as if integrating many nodes, a program controller node that has only a UECS-CCM communication function is necessary as a commander. The system works as a time-programmed multi-environment control system.

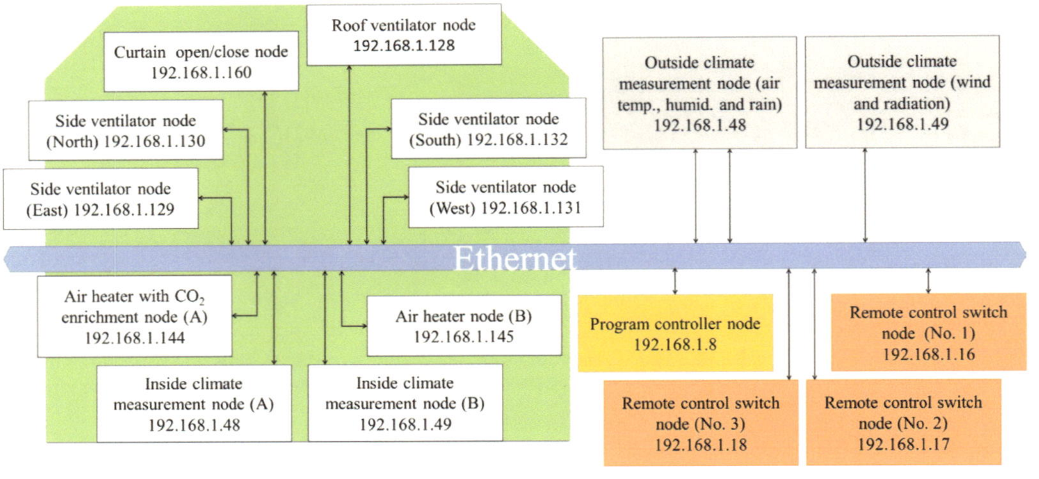

Figure 5. Configuration of the UECS nodes for the greenhouse used for tomato production [6].

The decentralized autonomous and cooperating environment control of all nodes by mechanisms of the UECS-CCM functioned satisfactorily to produce quality hydroponic tomatoes. The program controller node delivered UECS-CCMs for remote operation orders and control set points, and each node worked according to these. By capturing the UECS-CCMs on the LAN with a PC and storing them in a Comma Separated Value (CSV) format file, the environmental trends in the greenhouse and behavior of each UECS node are recorded easily.

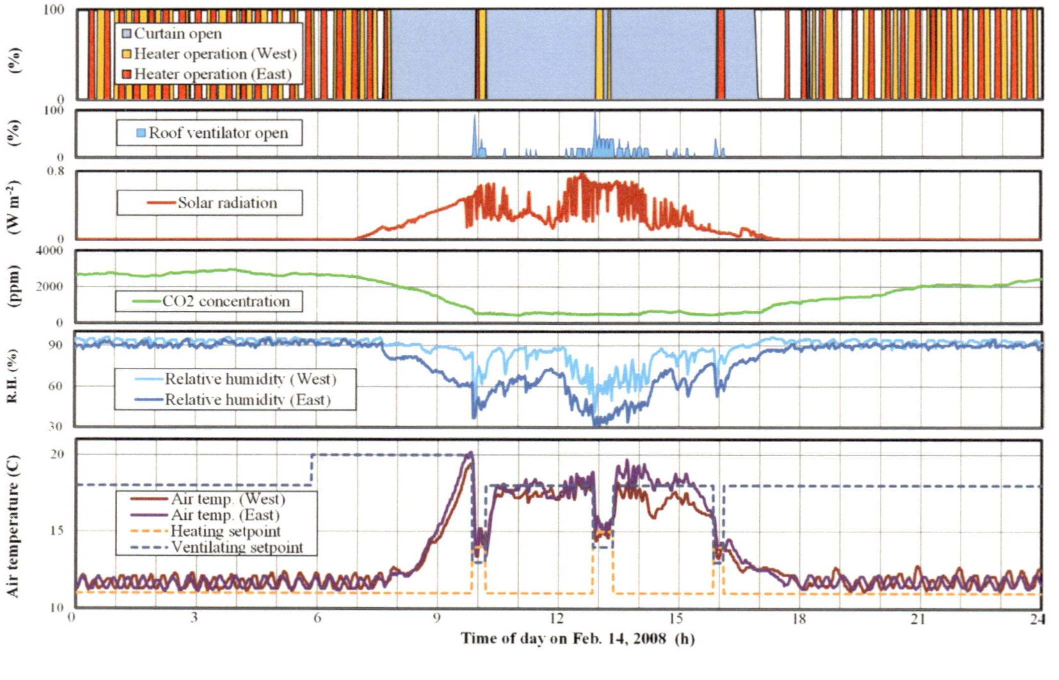

Figure 6. Time courses of measurement, control and set point values on February 14, 2008 of the installed UECS [6].

Figure 6 shows a chart using spreadsheet software for 1 day recorded in this manner. The dehumidification control for operation three times per day by cooperation of the air heater nodes and roof ventilator node was confirmed. If the program controller node stopped the CCM packet transmission due to failure or loss of power, all nodes returned automatically to the autonomic control state. The programmed multi-environment control system using the UECS was superior to the conventional control system in reliability.

Research and development of UECS technology in Japan are mainly being promoted by the "UECS consortium" established in 2006. Sales and installations of UECS products are handled by the "Smart-Agri Consortium," a consortium of companies that was founded in 2012.

3. DIY UECS for small-scale greenhouse growers

3.1. Why DIY (do-it-yourself)?

Small-scale greenhouse (less than 0.2 ha floor area) growers were not able to relish the benefits of the costly environmental control system previously. If the greenhouse growers could produce the equipment by themselves, they could easily get what they need. With a decentralized UECS, the environmental control system can be assembled step by step, so the difficulty of do-it-yourself (DIY) installation is relatively low compared to other systems. It is unnecessary to buy a costly control device in a small greenhouse. Growers will be able to improve and repair the equipment on their own. They do not need to spend a lot of money to employ experts. Making an environmental control system themselves, they take various advantages. Traditionally growers visually observed the state of the crop and manually controlled the cultivation environment. However, humans cannot keep observing plants for days without rest. If they could easily install sensors in the greenhouse, the understanding of the crop would be much deeper. However, until now small-scale greenhouse growers have had no technology they could install independently even if there was a control method they wanted to use. DIY UECS could provide a control unit that can be programmed as desired for greenhouse growers.

3.2. Low-cost microcontroller boards opened up DIY's way

The UECS system is decentralized, and it is necessary to incorporate a microcomputer in all the equipment in the greenhouse for communication and control (**Figure 7**). The educational microcomputer boards are inexpensive enough to have no problem even if it is built in all the equipment in the greenhouse, it is a promising platform which is mature (seasoned) and easy to obtain.

In recent years, two kinds of educational microcomputer boards called Arduino [8] and Raspberry Pi [9] have been widely used in various fields. Arduino was developed by an Italian manufacturer in 2005. It has many variations, but the price of the Arduino UNO (basic model) is around $25. Raspberry Pi was developed later and released in 2012 and is a microcomputer board with a higher performance than Arduino. The performance is comparable to a small PC. The price is also cheap at around $35. These microcomputer boards have acquired a large number of users and have been used not only for the educational and hobby applications originally developed for but also for industrial use [10].

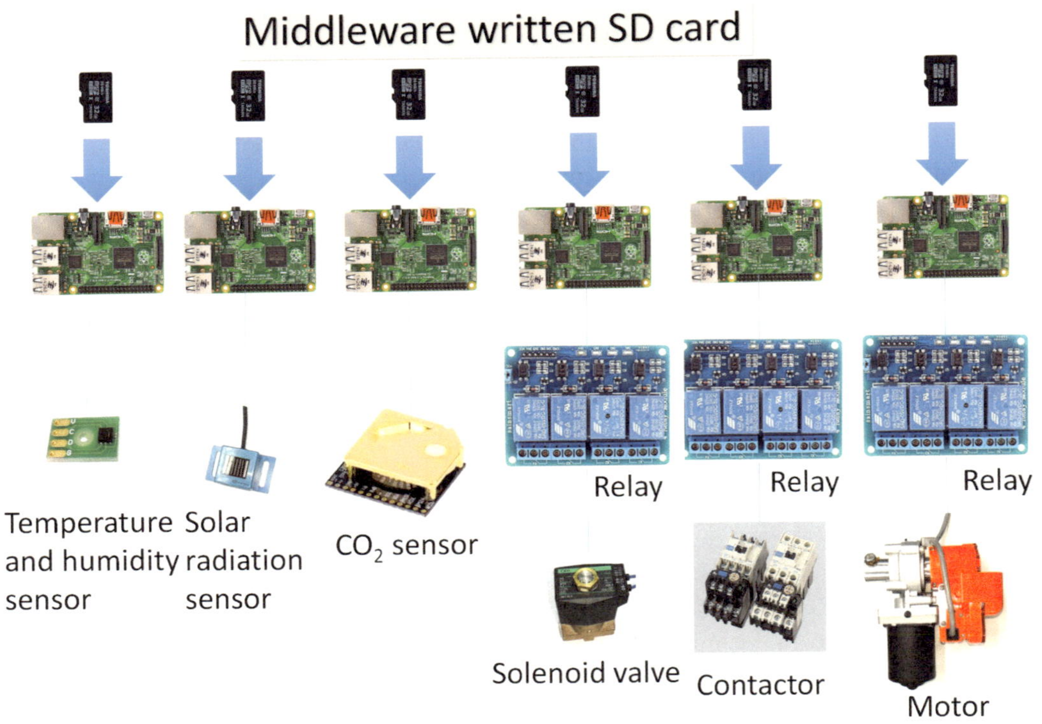

Figure 7. High versatility of microcomputer boards (e.g. with Raspberry Pi). Various UECS nodes can be generated with only a few parts exchanged. According to our experience, the SD card used for Raspberry Pi needs to be a solid industrial SD card to prevent breakage due to long-term use.

3.3. Software to support the development of UECS nodes

A program for mutually transmitting and receiving the UECS-CCM between nodes is necessary in order for each microcomputer board to function as the UECS node. The development of the implementation program is a difficult task even if you have programming knowledge and experience. Therefore, a library and middleware to implement the UECS-CCM were developed to enable the UECS node to be fabricated on a DIY basis.

For Arduino, the software, UARDECS, was developed in 2013, and is a library for incorporation into Arduino IDE, which is the official Arduino development tool and supports developers implementing UECS protocol and simple Web server function in Arduino. It is written in C language and requires knowledge of programming in order to use it. However, the advantage of using Arduino is that there are so many variations on the supporting device. When using Arduino UNO, you can create simple nodes with one or two sensors. When installing more devices, Arduino MEGA with a large memory load is suitable. UARDECS has already begun to be used for in university education and there are plans for it to be used in commercial greenhouses after beta testing. UARDECS will be released free of charge [11], and the development of the program will be carried out by universities, research institutes, or users with the technology.

Middleware for easily adapting the Raspberry Pi series to a UECS was developed by WaBit Inc., and the basic model "UECS-Pi Basic" is distributed as free software [12]. When Raspberry Pi is booted from the SD card in which the UECS-Pi is written, the Web server starts up. After that, if you access Raspberry Pi from a PC and so on, any customization can be done with using the browser-based interface. UECS-Pi is an extremely versatile tool that can be used for both sensing and control. The corresponding device is limited to those specified in the instruction manual, but its number is increasing due to version upgrades. At the time of this writing, available sensors range from temperature, humidity, CO_2, digital pulse, analog voltage, visible light camera, and thermal imaging camera. Functions for control mounted on the UECS-Pi can turn a switch ON/OFF and change the operation of the actuator based on the conditions entered by the user. UECS-Pi can be used by people who do not have any programming knowledge.

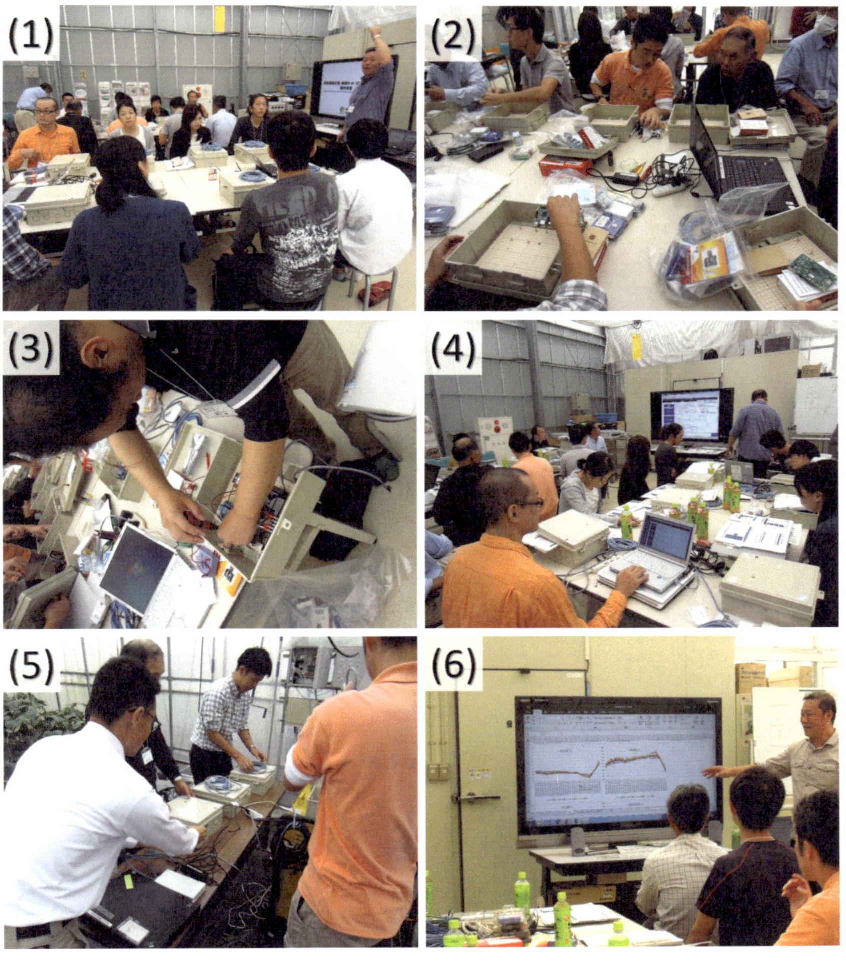

Figure 8. Snapshots of a workshop for self-manufacturing an UECS node. (1) Distribution and explanation of parts, (2) parts installation, (3) wiring connection, (4) set up by accessing from PC, (5) the completed node placed in the greenhouse, and (6) discussion on collected data.

3.4. Hosting self-made workshops for UECS node manufacture

As the groundwork for individuals to manufacture nodes for UECS has advanced, self-made workshops for UECS nodes have been held. As an example, the workshops held at Tsukuba's Institute of Vegetable and Floriculture Science, NARO in October 2016 (**Figure 8**) are introduced. There were 20 participants in the workshop, mainly farmers and staff of agricultural equipment manufacturers. The node to be created was the one using UECS-Pi based on Raspberry Pi. First, a set of parts and a manual were distributed. Participants installed microcomputer boards, power supply units, sensors, and other components and made wiring. After assembling was completed (**Figure 9**), they turned on, accessed from PC and set up. After confirming that the manufactured node operated normally on the desk, we brought it to the greenhouse and started measuring temperature and humidity. A data log was collected the next morning, and the recorded information was discussed. While some participants misinterpreted the wiring in the middle, everyone was able to finish the node at the end.

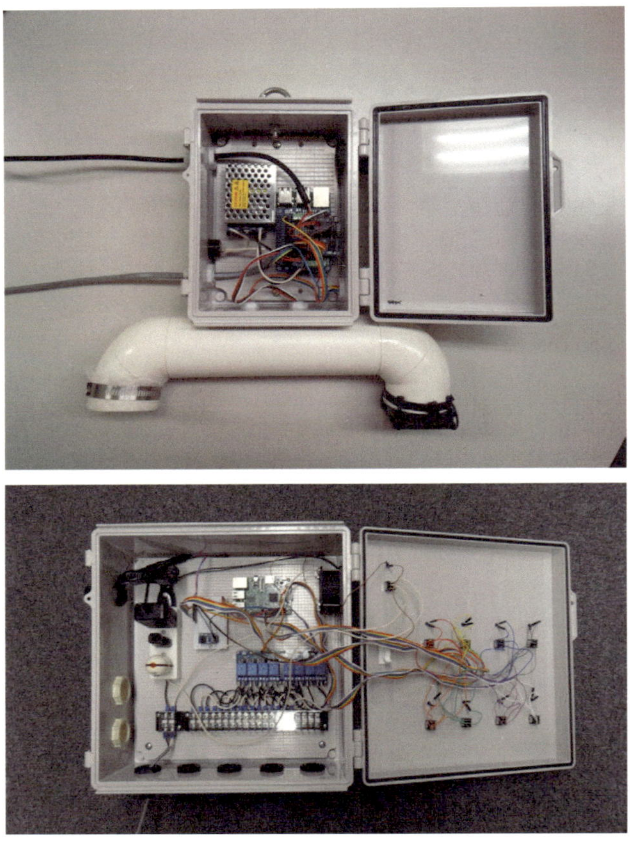

Figure 9. The UECS DIY node completed by the workshop. Upper: sensor node for measuring greenhouse climate, lower: relay node for controlling actuators such as open-close motor of roof ventilators.

http://dx.doi.org/10.5772/intechopen.71661

DIY brings new possibilities to horticultural production. Greenhouse growers familiar with plants may be able to create completely new, sophisticated cultivation techniques with their own hands. DIY is a powerful means for greenhouse growers to evolve horticulture with their own hands.

4. Upgrading traditional environment controllers using ICT

4.1. Upgrading traditional controllers using ICT

It is difficult to buy an environment controller with both high performance and low cost. Therefore, we tried to renew or upgrade conventional controllers using ICT. Handling of collected data will be easy using ICT. A key aspect for installing ICT into greenhouse production is the development of a controller conforming to the UECS. In this section, the advantages of renewing existing conventional controllers using ICT are discussed.

4.2. Configuration of renewed controllers

A renewed controller is composed of the conventional multienvironmental controller (base controller), microcomputer, Ethernet cable, and personal computer (**Figure 10**). The commercial

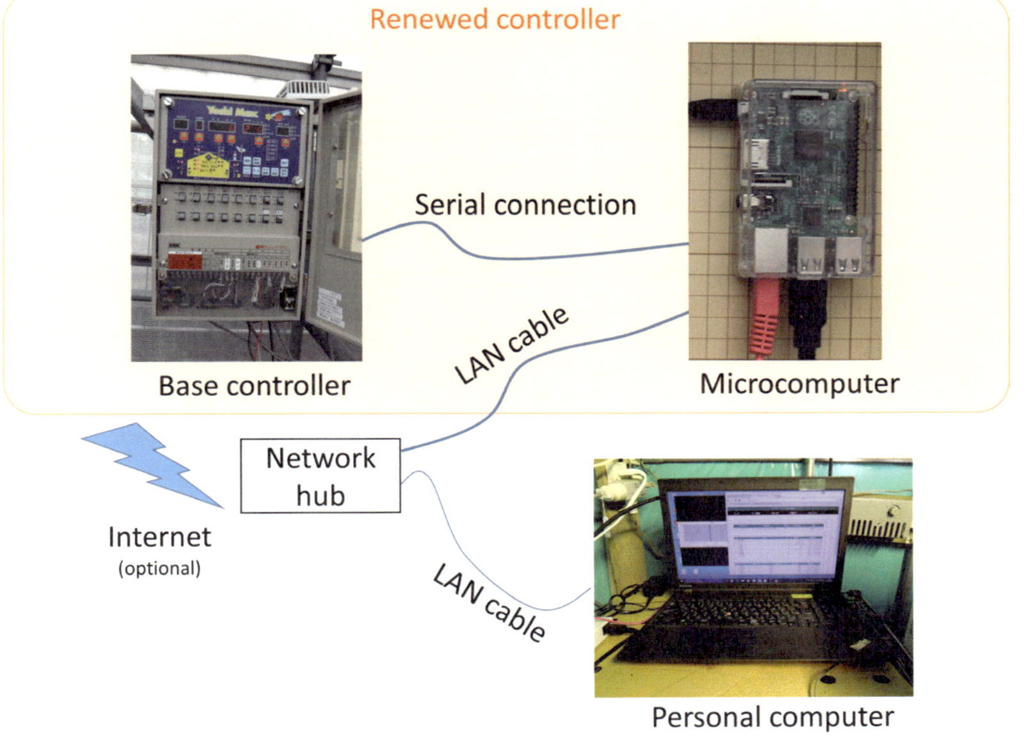

Figure 10. Configuration of renewed controller.

greenhouse environmental controller "Super-mini" sold by Sankikeiso Co. Ltd. chosen as one of the base controllers has already been installed in more than 100 greenhouses in Japan. It can be connected to the UECS node implemented by a microcomputer (e.g. Raspberry Pi) with a serial communication cable. The base controller frequently measures environmental conditions such as air temperature, solar radiation, and CO_2. All the measured data are directly transferred by the UECS-CCM to the other UECS nodes through the microcomputer. The microcomputer gives instructions for various controlling devices to the base controller after analyzing environmental data. The base controller can control connected devices using the direction of microcomputer. The HTTP server works in the microcomputer. Users must access the website using Web browser software to confirm sensor data and conditions of devices and to set parameters for environmental control.

The controller consists of three key devices. The use of a base controller is effective for the development of renewed controller speed and stability. Hardware design for environmental measurement and control operation of the base controller will be improved continuously, and it becomes robust now. Therefore, the development time and cost of hardware will be decreased. The microcomputer is necessary for controlling the ideal environment control and constructing a user interface for setting parameters for control. It therefore becomes possible to comply with a UECS using a microcomputer. A PC was used for accessing a microcomputer with a browser and monitoring information sent from the microcomputer with data collection software for UECS [13].

4.3. Case study on the development to use for strawberry production

In Okayama University, the environmental control technique for cultivating strawberry with high bed has been the subject of research for the past 10 years. The developed logic had been operated by using both the MS-DOS computer program and controller of end of sales for greenhouse in the previous system. If either this computer or the controller is broken, the developed logic could not be continued. Therefore, we renewed the controller as mentioned in the previous section. The controller named "YoshiMax" has several features as follows:

(1) CO_2 concentration can be flexibly controlled with a generator of CO_2 (**Figure 11**). The optimum level of CO_2 concentration can be changed by controlling the air temperature inside the greenhouse. If the ventilator is open, the generator is stopped. CO_2 generator of fuel combustion type could be used for air heater. The control of CO_2 concentration becomes easy, and the CO_2 generator was used effectively by this controller.

(2) It is possible to irrigate an amount proportional to the amount of solar radiation.

(3) The convenience of the UECS can be experienced by using this controller. Monitoring the environment can be easily constructed using data logging software for a UECS. If other UECS nodes were introduced previously, the environmental information can be gathered together with that on the controller. Users can set the parameters for accessing the controller using a Web browser.

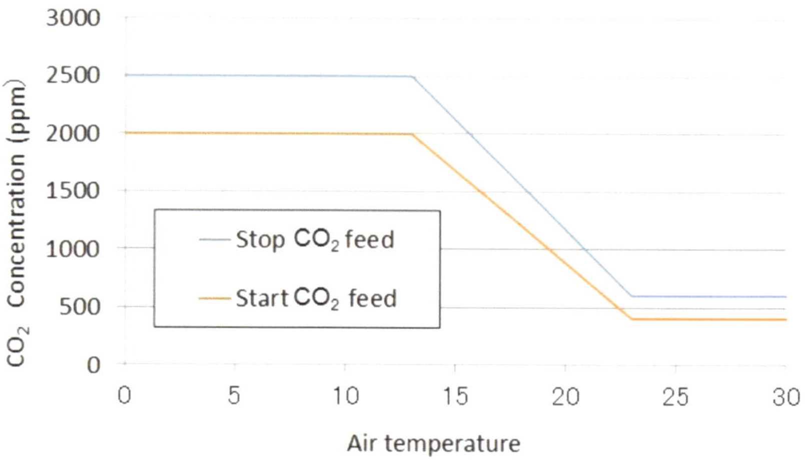

Figure 11. Control strategy for CO_2 gas enrichment suppressing CO_2 waste due to ventilation. The graph illustrates the relationship between the air temperature and control set point of CO_2 concentration.

Implementation of such flexible control algorithms and acquiring detailed performance information on each measurement control instrument cannot be realized only with the conventional controller. We have demonstrated that conventional controllers updated with UECS extensions are equivalent to new installations of the latest UECS compliant controller. Now, we have tested the controller with various strawberry growers and are working on widespread use of controllers.

4.4. Renewal controller to make introduction of ICT and the UECS easier

Making controllers compliant with the UECS was useful for the easy introduction of ICT. There are two main advantages of introducing a UECS. One is the standardization of greenhouse production information. The previously developed data logging software [13] can be used for logging data, and other nodes could be managed on the same system. Another is that LAN is introduced into the greenhouse. UECS is a system that carries out environmental control using LAN. Therefore, it is easy to communicate information in the greenhouse by using the Internet and to know current greenhouse environmental status by using a network camera.

Systems that comply with an UECS can be DIY-made. Although constructing a DIY system means it can be constructed cheaply, expertise on the hardware needed is also necessary. Modifying conventional controllers means that it is possible to develop equipment quickly without developing such hardware. Many users will prefer controllers that hate mechanical risks of hardware design and operate stably. It is considered that remodeling conventional controllers with a UECS is a useful technique for constructing a controller that operates with new logic quickly.

5. Wireless network applications for constructing a virtual large-scale horticultural farm

5.1. Limitations of a Wi-Fi network

In Japan, typical growers own several greenhouses that are distributed at a prescribed distance. Management of the greenhouses is performed individually and is therefore complicated and costly. To avoid this problem, a network infrastructure for distributed greenhouses to establish a virtual large-scale horticultural farm is required. For instance, a wireless network connecting each greenhouse would be the most effective network and has advantages such as simple installation and rearrangement of nodes, and a reduction of instillation time and cost. As mentioned in Section 2.2, the UECS has fully supported Wi-Fi network without any modification and addition of the system software and hardware in principle. However, the actual application in agriculture has several problems as follows:

(1) The Wi-Fi network is a de facto standard wireless local area network. The transmission quality can suddenly drop due to radio interference between routers if a lot of Wi-Fi routers are in use near greenhouses in a residential quarter owing to radio band conflicts.

(2) Wi-Fi commonly uses 2.4 and 5 GHz radio bands in Japan. The transmission accuracy given by throughput and packet loss remarkably depends on the environmental condition in the greenhouse. In particular, the radio wave is attenuated by vegetation and humidity conditions as the plants contain much water and discharge vapor by transpiration.

Owing to the above-mentioned reasons, the application of Wi-Fi networks in agriculture is limited only to cases where each UECS node is distributed close to a Wi-Fi router. Therefore, we concluded that a Wi-Fi network is not suitable to unify a distributed greenhouse network, that is, to establish a virtual large-scale horticultural farm.

5.2. Wide-area network using Sub-GHz radio bands

Recently, various types of wireless network standards and protocols have been developed and used as a fundamental network infrastructure in our society. Among them, Low-Power Wide-Area Network (LPWAN) has features such as a long transmission distance (max 50 km), several radio frequencies (typical radio frequency is 920 MHz called Sub-GHz radio bands), low transmission rate (max 250 kbps), and several standards such as LoRa, Wi-SUN, and IM920, which have no interchangeability in general. These features agree well with the requirements to establish a network covering distributed greenhouses for constructing a virtual large-scale horticultural farm.

Figure 12 shows an example of LPWAN applied for distributed greenhouses. In actual greenhouse management, growers expect to understand the current environmental information such as air temperature, humidity, and CO_2, and the control and configuration information for facilities installed in the greenhouse anytime and anywhere. The gateway in each greenhouse is able to correct the information in the greenhouse by using UECS-CCM. Here,

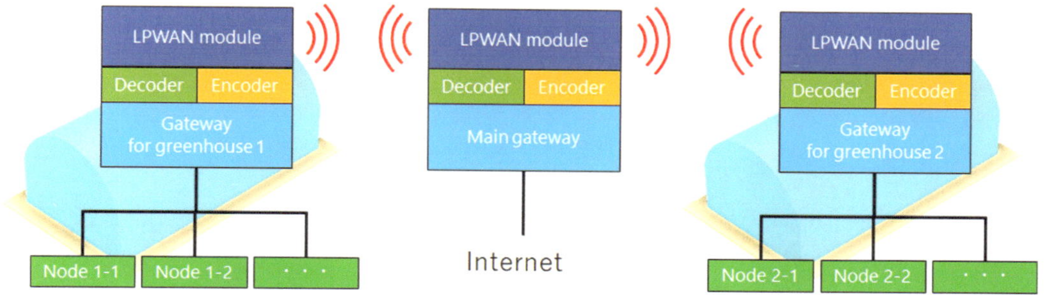

Figure 12. An example of LPWAN applied for the distributed greenhouses.

the original UECS-CCM is formatted using XML while the message size becomes large in general. Encoding to a transmission message based on LPWAN and decoding to UECS-CCM are required to reduce the size of the transmission message using LPWAN since the transmission rate is limited. Moreover, the transmission interval and message size are managed by the gateway in the greenhouse. The information is transferred directly to the main gateway for the Internet and is stored in the database. Therefore, the growers can review the information in all the greenhouses by using PC and mobile devices such as mobile phones, tablets, and mobile game consoles. At present, the authors are developing the prototype nodes based on this concept, for example, environmental monitoring node, data collection, and transmission node.

6. Conclusion and future perspective

When we proposed adopting the Internet protocol for the decentralized environmental control system for greenhouse in 2004, negative suggestions were obtained from many researchers and engineers in terms of reliability and real-time communication. However, in recent years, the term IoT has been popular, and the interest of UECS is increasing. In Japanese agriculture, facing to aging of people and farm land declines, UECS, which promotes automation and efficiency of protected horticultural production, is attracting attention as one of the important technologies for achieving safe and sustainable food production.

The improvement of greenhouse crop production in Japan is progressing in two major directions. They are (1) to construct a new large-scale and well-equipped greenhouse reflecting the types existing in Europe and the USA and (2) to install ICT systems for upgrading productivity in existing small- to medium-scale greenhouses. Our current research and development of UECS is proceeding with the aim of the latter. Therefore, issues of DIY, renewal, and wireless communication between discrete greenhouses have been the primary themes of UECS research and development, and various results have been obtained as mentioned earlier. These achievements are being adopted by greenhouses in six prefectures in Japan in the field test project "Realization of smart-agriculture by UECS platform to activate Japanese type greenhouses," which began in 2016. Upon completion of the project in 2019, the effectiveness of the UECS in agricultural greenhouses realizing low-cost and high-productivity greenhouse crop production will have been demonstrated and further spread will be expected.

Acknowledgements

This research was supported by grants from the Japan Society for the Promotion of Science KAKENHI Grant Number: JP25292157 and the Project of the NARO Bio-oriented Technology Research Advancement Institution (the special scheme project on regional developing strategy). We also would like to express our gratitude to the UECS consortium for assisting.

Author details

Takehiko Hoshi[1]*, Ken-Ichiro Yasuba[2], Hideto Kurosaki[3] and Takashi Okayasu[4]

*Address all correspondence to: hoshi@hoshi-lab.net

1 Faculty of Biology-Oriented Science and Technology, Kindai University, Wakayama, Japan

2 Graduate School of Environmental and Life Science, Okayama University, Okayama, Japan

3 NARO Western Japan Agricultural Research Center, Kagawa, Japan

4 Faculty of Agriculture, Kyushu University, Fukuoka, Japan

References

[1] Takakura T, Kozai T, Tachibana K, Jordan KA. Direct digital control of plant growth. I. Design and operation of the system. Transactions of the ASAE. 1974;**17**(6):1150-1154

[2] Zeeland J, Asselt CJ. A portable low cost microcomputer system for glasshouse climate control. Acta Horticalturae. 1981;**115**:181-190

[3] Sakamura K. The Tron project. IEEE Micro. 1987;**7**(2):8-14. DOI: 10.1109/MM.1987.304835

[4] Weiser M. Some computer science issues in ubiquitous computing. Communications of the ACM. 1993;**36**(7):75-84

[5] Hoshi T, Hayashi Y, Uchino H. Development of a decentralized, autonomous greenhouse environment control system in a ubiquitous computing and Internet environment. In: Proceedings of the Joint AFITA/WCCA, August 9-12, 2004; Bangkok, Thailand. 2004. p. 490-495

[6] Hoshi T, Hayashi Y, Shintani K. A communication protocol for collaboration among the measurement and control nodes in a decentralized autonomous environment control

system of greenhouses. In: Proceedings of IAALD/AFITA/WCCA 2008; August 24-27, 2008; Tokyo University of Agriculture, Fujisawa, Japan: 2008. p. 127-134

[7] Members of the UECS-CCM Standardization Committee, the UECS Consortium. UECS standard communication protocol books (UECS-CCM) [Internet]. June 9, 2012. Available from: https://www.uecs.jp/en/protocol-e.html [Accessed: September 7, 2017]

[8] Arduino. Arduino-Home. Available from: https://www.arduino.cc/ [Accessed: September 12, 2017]

[9] Raspberry Pi Foundation. Raspberry Pi—Teach, Learn, and Make with Raspberry Pi [Internet]. Available from: https://www.raspberrypi.org/ [Accessed: September 12, 2017]

[10] Designspark. Raspberry PI and Arduino in Industrial Environments [Internet]. April 29, 2016. Available from: https://www.rs-online.com/designspark/raspberry-pi-and-arduino-in-industrial-environments [Accessed: Septemver 12, 2017]

[11] Low-cost UECS Group. Download Page for UARDECS (written in Japanese) [Internet]. April 2013. Available from: http://uecs.org/arduino/uardecs.html [Accessed: September 12, 2017]

[12] Wabit Inc. Special Site for Smart-Agri Business (written in Japanese) [Internet]. Available from: http://www.wa-bit.com/smart-agri-project/product/uecs-pi-basic/ [Accessed: Spetember 12, 2017]

[13] Yasuba K, Kurosaki T, Takaichi M, Suzuki K. Development of management software for protected horticulture based on the practical protocol of ubiquitous environment control system. Bulletin of the National Institute of Vegetable and Tea Science. 2012;**11**:63-72

Chapter 7

Machine Vision Systems – A Tool for Automatic Color Analysis in Agriculture

Ernesto Martínez Sandoval,
Miguel Enrique Martínez Rosas,
Jesús Raúl Martínez Sandoval,
Manuel Moises Miranda Velasco and
Humberto Cervantes De Ávila

Additional information is available at the end of the chapter

http://dx.doi.org/10.5772/intechopen.71935

Abstract

It was in the early 1960s when machine vision systems initiated researchers and developers have worked on building machines that perform tasks of acquisition, processing, and analysis of images in a wide range of applications for different areas. Currently, along with the new technological advances in electronics, computer systems, image processing, pattern recognition, and mechatronics, it has arose the opportunity to improve machine vision systems development with affordable implementations at lower cost. A machine vision system is the combination of several high-tech techniques, including both hardware and software, used to acquire, process, and analyze images on a machine, which contributes with a set of tools for the extraction of features, such as color and dimension parameters, texture, chemical components, disease detection, freshness, assessment, modeling, and control, among others. Based on former paragraphs, we could say that machine vision systems are appropriate to improve the actual agricultural systems making them more useful, efficient, practical, and reliable.

Keywords: image processing, computer vision, color analysis, habanero chili, vineyards

1. Introduction

The evaluation of what surrounds us is done through light, colors, shapes, textures, and intensities, among other characteristics, which originate from different natural phenomena that give rise to spectacular scenes that are visually striking to observe. The way to perceive such an enormous amount of information is achieved through the senses. The perception of the human being is an incredible skill composed of five senses: sight, ear, smell, taste, and touch. The human senses are complex systems within the human body, which in turn are formed by an immense number of well-adapted and calibrated sensors with a very wide range of operation to carry out specific tasks, in addition to adapting and interacting with each other. Some of the senses manage to partially or totally regenerate from time to time. The human being has another essential organ, the brain, which is the main processing unit, responsible for receiving, processing all information, making decisions, and coordinating actions in a synchronized, a fast, and an efficient manner. That is why trying to emulate these capabilities is a huge challenge.

The visual perception of the human being is basically composed by the interaction between the eyes and brain. In general, the human vision system can be described as follows: the eyes are composed of the eyeball and the muscles that control its position. The cornea and lens focus light rays at the back of the eye. The lens regulates the focus for near and far objects as they become more or less globular. On the other hand, the brain is much more complex. Even to this day, there are many unknowns about its operation and even its biological purpose. Computer vision systems (CVS) intend to emulate this sophisticated and dynamic vision system whose operation is very natural and transparent [1].

Computer vision can be understood as the science that develops theory and algorithms to extract useful information about an object or scene within an image, for further analysis. Computer vision systems follow a basic structure that can be divided into three stages illustrated in **Figure 1**: low, intermediate, and high level of processing. In the first stage, the appropriate image acquisition sensors must be chosen for the type of scene to be

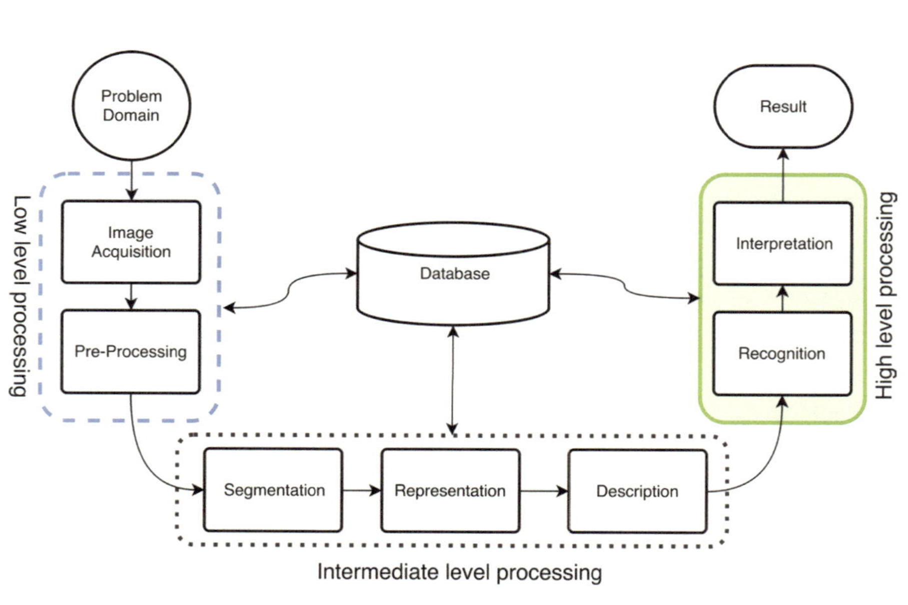

Figure 1. Stages of the basic structure in which image processing is divided.

captured, as well as to establish the acquisition conditions that best fit the experiment in order to obtain clear and good quality images. For the second stage, we must work on segmentation, which is considered a fundamental process in the image processing. The segmentation process separates useful information from the rest of the scene; for this, there are different segmenting and filtering techniques whose objective is to define and extract desired features within the image. The third stage, of high level performs an analysis with the information resulting from the two previous phases, so that its performance is closely linked. This analysis is focused on the recognition of objects within the scene and interprets their attributes to give results or make decisions. This is based on the information extracted and a knowledge database that serves to store the information of the whole process, as well as to train the vision system. Image processing and image analysis are the core of computer vision systems.

Both human beings and systems of vision rely on light to appreciate their surroundings. Therefore, the sources of illumination as well as its quality are crucial parameters for the proper functioning of both biological and artificial systems. Roughly speaking, light is the portion of the electromagnetic spectrum that can be perceived by the human eye (400–800 nm). The reflections of different wavelengths in materials are essentially what we call the color of an object. Depending on the physical characteristics of the materials, light incident on an object can give rise to three phenomena: absorption, when light on the object converts that energy into heat; refraction, when light passes through the material by changing its direction as a function of the refractive index of the material; and reflection, when the material rejects certain wavelengths. The reflected waves are those that can be observed with the naked eye and give rise to the familiar range of colors.

Color is the result of the interaction of the light reflected by the objects upon contact with the cells of the cones of the human vision system. There are different kinds of cones in the human eye, so the level of absorption and the interpretation of the nervous system of these signals give rise to the perception of color. So, it can be said that color is a perception of light reflected by the surface of an object [2].

In computer vision systems, the color attribute is very valuable and is very often used to identify useful information and perform its extraction. It should be mentioned that one of the main factors in defining the quality of products in agro-industry, particularly fruit and vegetables, is their appearance and, therefore, the value of these products in the market. Appearance is a combination of color, shape, size, and texture, among others. As a result, color makes the product very attractive to consumers and helps them buy the product.

2. Machine vision systems

From a general perspective, machine vision systems (MVS) or computer vision systems (CVS) try to emulate human vision in order to gather information from an object without requiring a physical interaction with it [3]. Machine vision is a complex high-tech system which includes an image sensor (usually a charge-coupled device, known as CCD), a frame grabber and a computer with the appropriate software and algorithms [4]. In general, the procedure is as follows: a scene is captured by the image sensor, the analog electrical signal obtained from the sensor is converted to a digital format and sent to a computer, and then it is subsequently processed using algorithms in order to analyze the corresponding image. This procedure is frequently implemented using different setups to accomplish specific application requirements. The importance of this technology for the analysis of agricultural products lies in its nondestructive evaluation characteristic.

Among the several definitions for CVS offered by different authors, Timmermans says that CVS are states that include capture, processing, and analysis of 2D images [5]. Sonka mentions that the objective is to replicate human vision by perceiving and understanding an image electronically [6]. Jha holds that the perception of an object and their optical characteristics in order to perform an interpretation of the results is called vision system [7]. Also, it is defined as a system for automatic acquisition and analysis of images to obtain desired data for interpreting or controlling an activity. The system consists of image acquisition, image processing, and interpretation [8].

CVS are suitable for agricultural applications because when used to obtain the characteristics of fruits and vegetables the task is done quickly, economically, hygienically, consistently, and objectively. This is the reason why the use of CVS has been expanded in many sectors of the industry, such as medical, automation, surveillance, remote sensing, autonomous vehicles, and robot vision, among others [9–12].

There is a broad range of applications for CVS, ranging from routine inspection through complex vision systems for robots, which shows the flexibility of this technology. On the other hand, its implementation requires a relatively low cost, which makes this technology even more attractive for other applications.

In the case of the food industry, CVS have proved to be an alternative method for inspection of visual attributes in pastries [13, 14], meat [15–17], and fish [18, 19]. Because this inspection method is nondestructive, it is widely used in agricultural applications, including inspection and selection of fruits and vegetables [20]. For the agro-industry, the visual aspect of its products is particularly important, since this parameter is a determining factor of the value of the product in the market.

Traditionally, personnel trained for this task carry out the inspection and selection of agricultural products manually. This implies several disadvantages, such as inconsistencies in selection, time consumption, intensive tasks, variability, and subjectivity. In addition, the manual process is tedious, laborious, costly, and strongly dependent on external factors.

On the other hand, the appearance of agricultural products, i.e., their size, shape, and color, and the presence of stains or shocks, have a negative influence in the consumer perception and therefore determine the degree of acceptance prior to a purchase. The consumer also associates a certain internal quality with external characteristics (the appearance), which affects future decisions in the purchases [21].

Quality is a key factor that Kramer defines as:

http://dx.doi.org/10.5772/intechopen.71935

"Quality of foods may be defined as the composite of those characteristics that differentiate individual units of a product, and have significance in determining the degree of acceptability of that unit to the user" [22].

In reference to fruits and vegetables, these attributes can be classified as follows:

- Color and appearance are the factors that normally determine whether the product is accepted or not.
- Flavor (taste and aroma): once the product's appearance convinces the consumer and the product is tested, the flavors and aromas become more important. Freshness, itching, and sweetness are some of the attributes that can be detected when consuming certain products.
- Texture can be perceived externally not only when taking the product in our hands but also when you taste it, you have a clear impression of the softness, firmness, or crisp of the fruit or vegetable.
- Nutritional value is a factor that is usually hidden, but it affects our organism in ways we cannot perceive. However, it is an increasingly important parameter for consumers, scientists, and medical personnel.

These factors are closely related to the process and maturity stage of both the plant and the fruit at the time of harvest, as well as postharvest management conditions.

Shewfelt suggests "A primary disadvantage of instrumental testing is that many instrumental measurements have little relevance to consumer acceptability and thus should never be used

Products	Applications	Technology/technic	Reference
Rice	Panicle length measuring	Dual-camera devices	[24]
Soybean	Detection of insect-damaged soybean	Hyperspectral transmittance image	[25]
Apples	Detection	RGB-D	[26]
Cauliflower	Extract structural parameters to assess the growth	RGB-D	[27]
Cherry tomatoes	Detection of cuticle defects	Hyperspectral fluorescence imagery	[28]
Date fruits	Determining viscoelastic characteristics of date fruits	CCD	[29]
Dried figs	Grading	CCD	[30]
Foliar	Disease spots	CCD	[31]
Orange	Occlusion recovery	CCD	[32]
Rapeseed varieties	Classification	CCD scanner	[33]

CCD: charge-coupled device; RGB-D: color depth camera.

Table 1. Machine vision systems in agriculture.

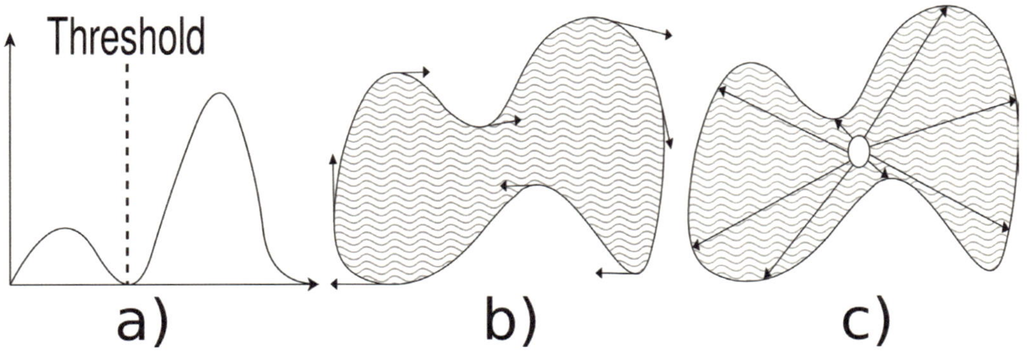

Figure 2. Typical segmentation techniques: (a) thresholding, (b) edge-based, and (c) region-based.

to define quality attributes for a specific product. In other words it is better to measure what is really important than to believe something is important because you measure it really well" [23].

Machine vision systems present a viable alternative to extract certain important attributes automatically and, moreover, analyze them for making decisions. Besides, they have been diversified in distinct areas that concern agriculture. Due to its wide range of use, a summarized categorization is presented in **Table 1**, considering the products that are evaluated, the type of application, and the technology that was employed.

The extraction of attributes of agricultural products presents important challenges due in part to the variety of products that exist. The extraction of useful data is highly dependent on a good segmentation. There are several techniques for segmenting images, three of which are illustrated in **Figure 2**. The first technique is thresholding, which is a fast technique that converts gray levels into a binary representation. The second technique is edge-based technique, which looks for change points that give way to borders or contours, and, finally, region-based technique, which performs a search around a group of pixels within the image.

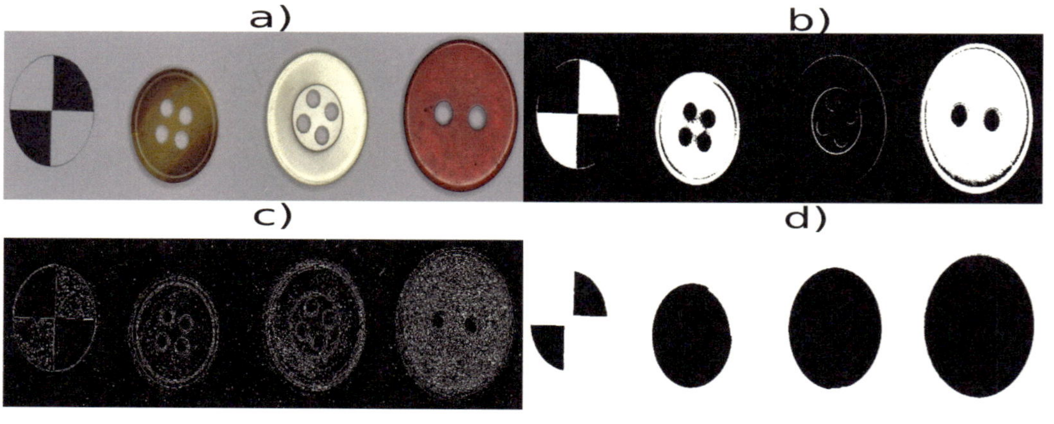

Figure 3. Applications of segmentation techniques: (a) original image, (b) thresholding, (c) edge-based, and (d) region-based.

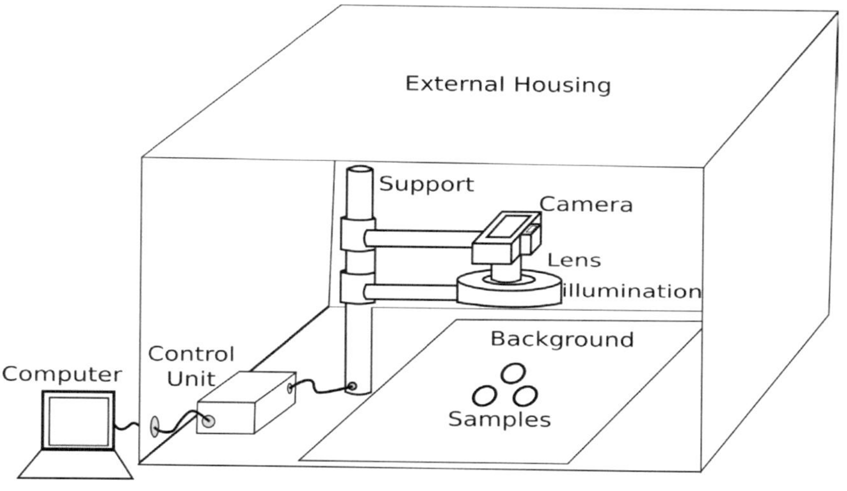

Figure 4. Typical machine vision system component.

To illustrate the performance of the different segmentation techniques described above, a desktop setup was implemented as shown in **Figure 4**. The first step is to capture an image; clothing buttons were used in this experiment, which also include a printed circular reference. The second step was to implement routines in MATLAB that process the image with the different segmentation techniques resulting binary images. The performance of the three techniques was satisfactory, but the processing time differed. Thresholding was the fastest, then edge-based and region-based. The resulting binary images are illustrated in **Figure 3**.

3. Hardware and software requirements

In order to acquire all the useful characteristics of the image, it is required to use a suitable light source to obtain the type and quality of illumination required over the scene. Taking into account that the image sensor will receive the light reflected from the scene, the upper level of illumination will be limited by the sensitivity of the sensor. The image sensor resolution must be taken into account, since it generates the amount of pixels from the image, where light information is contained, as it was perceived. On the other hand, the speed of frame grabber is selected considering the shortest period time of the task to be captured; for example, to acquire images on real-time or online applications, a faster frame grabber is required. Once the image is acquired, it is ready to be processed and analyzed; the algorithm used to perform the task is designed to extract specific parameters from the image and then to show them as results in the proper form.

The automatic operation of the setup involves to take into account timing considerations, to assembly high-tech components, to write proper and efficient program codes, and, finally, to synchronize and integrate every component in a single functional system. First of all, the scene capturing place, the proper illumination, and the required insulation conditions to acquire useful images must be selected. The final application of the machine vision system must be defined in order to get a suitable image sensor and a frame grabber that accomplishes the resolution and speed required. Finally, it is necessary to write program codes for three different stages on a machine vision system: to acquire, to process, and to analyze the images. Acquisition refers to the task where the image is taken. During the processing stage, the features are extracted, and the image is enhanced. Finally, once that parameters and features of the image are found, the algorithms of the analyzing stage use them to get a result in context to the goal of the application [34].

4. Color analysis

Our eyes are able to detect what we call visible light, which in fact is a range of electromagnetic spectrum wavelengths between 400 and 800 nanometers. What we observe a scene is actually the light reflected from the surfaces or the radiation emitted by light sources that make us to experience the sensation of color.

In reference to fruits and vegetables, color is derived from the natural pigments in fruits and vegetables, many of which change as the plant proceeds through maturation and ripening. The primary pigments imparting color quality are the fat-soluble chlorophylls (green) and carotenoids (yellow, orange, and red) and the water-soluble anthocyanins (red, blue), flavonoids (yellow), and betalains (red). In addition, enzymatic and nonenzymatic browning reactions may result in the formation of water-soluble brown-, gray-, and black-colored pigments. Color is one of the most important object measurements for image understanding and object description [35]. Using color attributes is more discriminant than simply using grayscale since two image points with the same grayscale value can be differentiated from their color attributes [36].

The level and quality of illumination from a light source affect the performance of the human eye, just as the performance of computer vision systems is affected by the illumination sources used. Sarkar found that by adjusting the illumination, the appearance of an object can be radically altered [37]. Therefore, the lighting system can greatly influence the quality of the images, so it plays an important role in the efficiency and accuracy of the CVS. Gunasekaran noted that a well-designed lighting system can help improve the performance of image analysis by improving contrast [38].

Some of the technologies used to acquire images in the food production sector are charge-coupled device (CCD) camera, magnetic resonance imaging (MRI), ultrasound, X-ray, near-infrared spectroscopy, and computed tomography (CT), among others. From the technologies mentioned, the CCD camera is widely used in computer vision systems for quality assessment, product selection, and product classification. CCD technology has the ability to convert captured light into electrical signals to create images. Depending on the characteristics of both the sensor and the optics used to acquire the images, these can be obtained with high resolution, low noise, and good light-sensitive (ISO) level. These characteristics can be adjusted up to certain limits to adapt the images to different capture scenarios. Both color and monochromatic cameras have been used in the food industry for a wide variety of applications [39–42]. There are several attributes used to evaluate the quality of products in the food industry. There are external attributes, such as shape, size, color, texture, and defects, which

Category	Products	Applications	Technology	Accuracy (%)	Reference
Fruit	Apple	Sorting by color and size	HSI	90%	[43]
	Apple	Mature discrimination	RGB	95.83%	[44]
	Starfruits	Color classification	HSI	100%	[45]
	Banana	Color measurement	RGB, HSV, L*a*b	97%	[46]
	Peach	Sorting by color and size	HSI	90%	[47]
Vegetable	Potato	Color classification	RGB	90%	[48]
	Potato	Blemish detection	RGB	89.6%	[49]
	Tomato	Color classification	RGB	No reported	[50]

HSI: color space (hue, saturation, intensity); HSV: color space (hue, saturation,value); RGB: color space (red, green, blue); CIELAB: L*a*b* color space.

Table 2. Analysis by color attributes.

can be captured by CCD-based vision systems. On the other hand, among the internal attributes of food products that may be measured, we can mention internal structures, quantity of water, and gaps. The extraction of these attributes requires the use of technologies such as ultrasound, MRI, and CT in order to obtain images with sufficient information for processing. From the point of view of image processing, both external and internal attributes present important challenges to adequately identify the information of interest. Moreover, each technology has different qualities, and its usefulness will depend on both the type of information to be extracted and the type of application. In addition, it is advisable to keep the economic factor in mind before deciding the type of technology to use.

The appropriate selection of the color space contributes to enhance color attributes from processed images. **Table 2** presents a simple classification of agro-products. The table consists of six categories: the first column helps to classify the products in fruits or vegetables; the second column presents the study applications; the third column has the technology or technique used; the fourth column has the percentage of efficiency reported; and the last column contains the references of each of the submitted works. In fruits, it is clearly noted that apples are a quite popular product and have been evaluated using different color spaces with an accuracy greater than 90%, as well as in banana and peach fruits. The applications range from sorting fruits by color and size, maturity discrimination, and color classification, to vegetables focusing on blemish detection and color measurements. The RGB, HSI, HSV, and $L^*a^*b^*$ color spaces are used depending on applications.

5. Applications of machine vision systems in agriculture

5.1. Machine vision system implementation

To acquire and analyze the information from an image, it is necessary to implement a machine vision system. The simplest case involves the implementation in a desktop setup. The primary goal of this configuration is to provide the insulation conditions to acquire a clean, high-contrast image. Additionally, in this implementation it is easy to add a number of sensors to monitor inside ambient conditions, in order to provide complementary data for the analysis. These characteristics are quite useful for the agricultural sector to characterize several species as well as a variety of crops under different conditions [51, 52].

A summary of technologies used in a wide range of implementations in the agriculture area is presented in **Table 3**. Products, applications, technologies (used as a machine vision system), the accuracy of this system, and the corresponding references are included.

Based on an extensive bibliographical research, the advantages and disadvantages of the artificial vision systems applied in agriculture are indicated in **Table 4**. It is important to mention that the nondestructive characteristic becomes one of its main advantages, since the industry wants to sell the final product, with the best quality to the consumer and in that way increase their profits.

5.2. Case study

5.2.1. Habanero chili color assessment

Setups for 2D vision systems are suitable for capturing images of fruits and vegetables. For this case of study, the setup required a camera hold steady at the center top for capturing the images. It was used a commercial digital single-lens reflex (DSLR) camera as capture device.

Products	Applications	Technology	Accuracy (%)	Reference
Oilseeds	Measuring thermal properties	MR	No reported	[53]
Passion fruit juice	Tracking thermal degradation	MR	No reported	[54]
Pear	Grading by external shape	Features	88.2%	[55]
Strawberry	Bruise detection	H-CVS	100%	[56]
Apple	Quality grading	M-CVS	93.5%	[57]
Apple	Chilling injury	H-CVS	98.4%	[58]
Citrus	Rottenness detection	H-CVS	98%	[59]
Mango	Mango grading	FD	89.83%	[60]
Mushroom	Enzymatic browning	H-CVS	>89%	[61]
Citrus	Defect detection	T-CVS	98.9%	[62]
Agricultural products	Internal characterization	X-ray	No reported	[63]
Agricultural products	Quarantine scanner	X-ray/CCD	No reported	[64]

T-CVS: traditional computer vision system; H-CVS: hyperspectral computer vision system; M-CVS: multispectral computer vision system; R: Magnetic resonance.

Table 3. Technologies applied to agriculture.

Advantages	Disadvantages
Consistent	Adaptable under specific conditions
Database	Artificial lighting
Efficient	Environment control
Fast	Object identification
Flexibility	
Non expensive	
Nondestructive	
Robust	

Table 4. Advantages and disadvantages of computer vision system.

As shown in **Figure 4**, the capture device includes a housing that allows to isolate the experiment from external contaminants such as light and dust. In addition, a control unit is installed to trigger the camera remotely in order to avoid handling vibrations and to ensure that the captures are taken as uniform as possible between samples. In the same way, the illumination lamp is turned on few seconds before to capture the image, which allows to stabilize the lamp and in this way to achieve a more uniform illumination.

To verify the proper performance of the capturing device, it is advisable to include a wireless communication to the control unit in order to perform a remote monitoring. On the other hand, the illumination lamp spreads the light using the white surface of the interior of the case, allowing a better distribution of light, reducing both shadows and reflections in the samples. Immediately, after each image capture, the lamp is turned off to avoid possible damages due to over exposure to the fruits. In addition, a high-contrast background has to be used and thus facilitate a correct segmentation.

To ensure that the conditions and positions of the samples do not change during the capture of the images that are same, each individual sample can be separated using specialized routines, so that they can be processed for further analysis. In order to extract the color attribute automatically from the images, a routine can be implemented to carry out a threshold segmentation and thus to obtain only the information of the region of interest (ROI). Once the information of the camera in RGB format is obtained, a color space conversion to CIELAB was made to show the color changes as a function of time.

Images were taken with a sampling frequency of one image per hour, during 8 days for 35 habanero chili specimens. This generates a total of 6720 sub-images from the 192 images taken. The type of images captured from this system can be seen in **Figure 5**.

The foundation for development of image processing algorithms was to keep them simple and with a proprietary processing instead of using toolbox functions of MATLAB® for image processing, except for the fundamental ones to loading and displaying images. The first step was to test whether the algorithms were able to capture, process, and analyze the color images

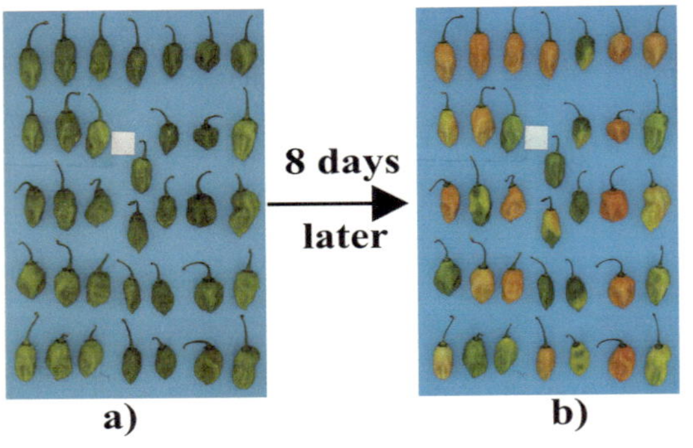

Figure 5. Type of images captured: (a) samples at the first day (hour 1) and (b) samples at the eighth day (hour 192).

of habanero chili. Since these algorithms are part of a larger project that considers migrating these codes to an embedded system, to provide greater versatility and portability to the vision system, expanding its range of applications. Algorithm starts loading an image, from color image database of habanero chili specimens from the previous acquisition stage. Then, a pre-determined area segmentation delimits a specific region for each specimen, called sub-image. Next, a color segmentation is performed, using a threshold technique, where the background color parameters are set as the threshold value, so everything else is the color information from the specimens. Then, a color space conversion from RGB to CIELAB is carried out. After that calculations of color average from each specimen are performed and color information in CIELAB components $L^*a^*b^*$, Cab^*, and hab° is generated, this stage results in a matrix with these color parameters. Finally, with this color data, a color analysis is executed, where these parameters can be plotted and observed, as shown in **Figure 6**.

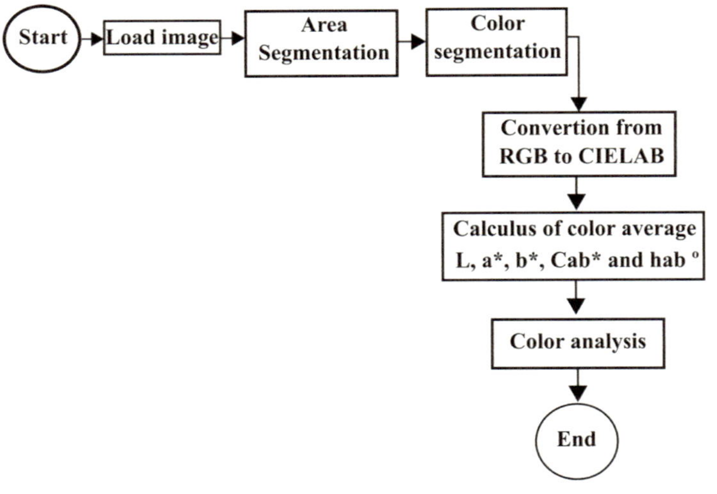

Figure 6. Flowchart of image processing algorithm.

http://dx.doi.org/10.5772/intechopen.71935

In **Figure 7**, a color diagram is presented with the parameters of a* vs. b* from CIELAB and represents the mean values from each specimen during 8 days of experiment. Each point signifies the average from 24 images taken per day, corresponding to sampling rate. It can be seen how the system is able to detect color changes.

Color information obtained from image processing and expressed in terms CIELAB coordinates was statistically analyzed applying a one-way analysis of variance (ANOVA). The analysis of variance shows highly significant differences between specimens. The obtained results are reported in **Table 5** (sum of square [SS], degrees of freedom [DF], mean square [MS], and test statistic [F]). Moreover, **Figure 8** represents the analysis of variance between specimens.

Another statistical analysis was performed using one-way analysis of variance (ANOVA). The analysis of variance shows highly significant differences between days. The obtained results are reported in **Table 6**. Moreover, **Figure 9** represents the analysis of variance between days.

Highlights: a color image database of habanero chilies at controlled conditions was generated for future developments and analysis. At the moment, developments in this specific kind of application explore noninvasive techniques for color assessment in habanero chili specimens; the information related was null, which increase the scientific value of these results. Algorithms developed are capable to acquire process and analyze habanero chili specimens in postharvest under controlled conditions. The guideline for algorithm development was to write code with basic functions and proprietary code, instead of toolbox functions of MATLAB®, considering code migration for future research and developments. The statistical analysis confirms that image processing algorithms and machine vision system are capable to detect, quantify, and analyze color changes of digital images from habanero chili in an automatic way.

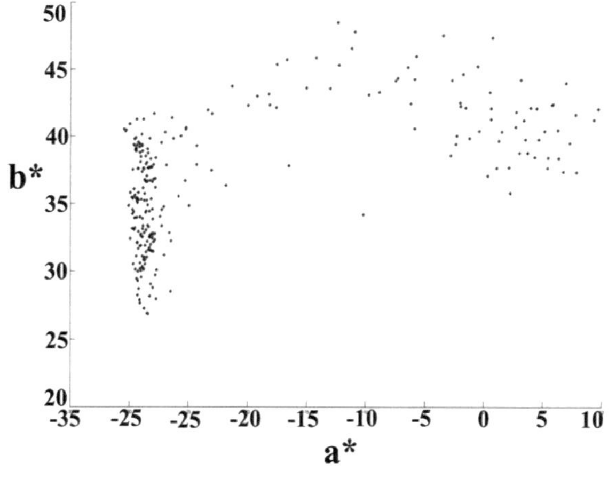

Figure 7. Color diagram of a* vs. b* of all samples from their mean values per day.

Source	SS	DF	MS	$F > F_{0.05}$
Days	22643.2	7	3234.74	10.97 ≫ 2.010
Error	80213.9	272	294.9	—
Total	102857.1	279	—	—

Table 5. One-way analysis of variance (ANOVA) between samples of $\alpha = 0.05$ significance.

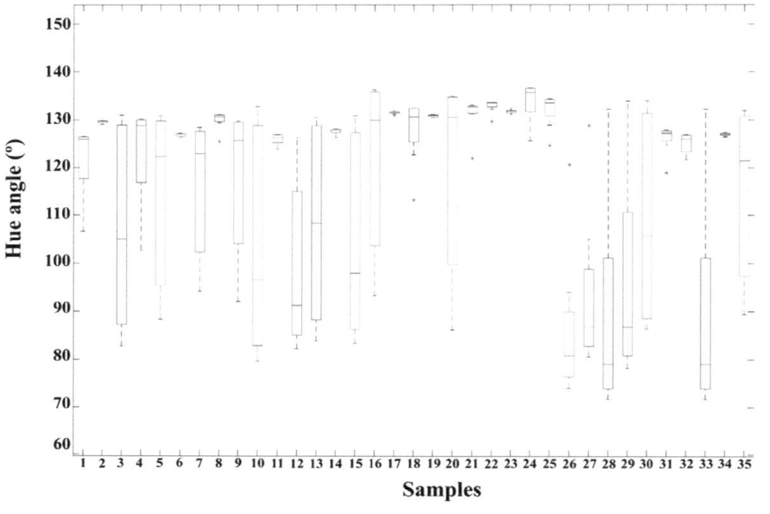

Figure 8. Analysis of variance results between samples, box, and whisker plot.

Source	SS	DF	MS	$F > F_{0.05}$
Specimens	56449.8	34	1660.29	8.77 ≫ 1.424
Error	46407.3	245	189.42	—
Total	102857.1	279	—	—

Table 6. One-way analysis of variance (ANOVA) between samples of $\alpha = 0.05$ significance.

5.2.2. Vineyards

The wine industry is one of the most interesting in the use of vision systems to increase the quality of its crops [65]. An increasing number of wine producers recognize the advantages of understanding the biophysical characteristics and the performance of their vineyards, leading to better management of their resources and making decisions. Wine producers commonly have a goal for the state of maturity they want to achieve for the wine they produce. Such a goal can vary, even within the same grape variety, depending on the type or style of wine that

http://dx.doi.org/10.5772/intechopen.71935

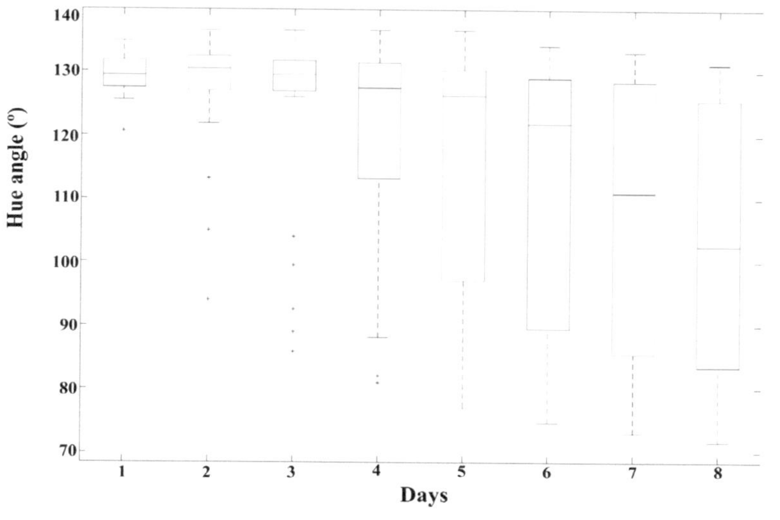

Figure 9. Analysis of variance results between days, box, and whisker plot.

will be made. As previously mentioned, for optimum determination of fruit ripeness, color is one of the most important parameters. In the case of vineyards, a computer vision system can be used to individually locate grapes in clusters as well as grape berries that are used in evaluation samplings. By using segmentation algorithms, sub-images of each grape are used to extract the information of the color parameters in a proper color space.

As mentioned by Martinez-Sandoval et al. [66] as well as Rabatel and Guizard [67], digital images of individual grapes could easily be acquired in the field as long as adequate control of the distance and light conditions over the samples is maintained. The use of computer vision systems would allow estimating the volume of grapes from their visible area, and using suitable image processing algorithms, it would be possible to obtain detailed information such as the color and size of the berries.

The development of methods based on image processing has advanced to the point of being able to automatically detect and count the grapes to accurately predict the yield of a harvest. As indicated by Nuske et al. [68], in order to capture the images of vines (in the visible light spectrum), conventional cameras are used throughout the vineyard. A typical algorithm to perform the image processing of vines can be divided into the following stages [66]:

1. Detecting potential berry locations with a radial symmetry transform.
2. Identifying the potential locations that have similar appearance to grape berries.
3. Group neighboring berries into clusters

A vine image consists of a set of berries with nearly identical color properties, so that the information available for berry separation mainly relies on their contours (as seen in **Figure 10**)

Figure 10. Image segmentation of bunch of grapes.

where a bunch of grapes are segmented from the background. Due to the smooth 3D shape of the berries, their contours are accessible as luminance discontinuities in the image. However, the algorithm only works with grayscale images. The main goal is to make a nondestructive estimation of the bunch weight of grapes at an early stage by computer vision using an elliptical model suitable to estimate the volume of the grape from its visible area.

Winemakers commonly have a goal for grape maturity that they would like to achieve for the wine they will produce. This objective can vary, even within the same grape variety, depending on the type or style of wine desired [+]. For the characterization of maturation, physicochemical analyses such as sugar content, acid, and pH [69] are performed, sugar content being one of the most important factors [70]. **Figure 11** shows a generalized graphical representation of changes in grape composition during development and ripening.

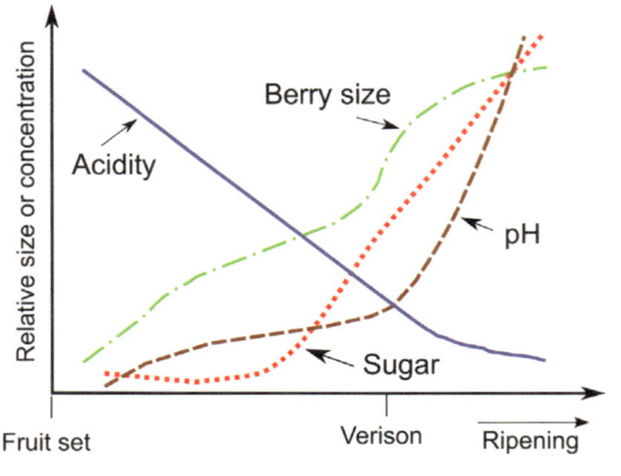

Figure 11. Grape berry development and ripening.

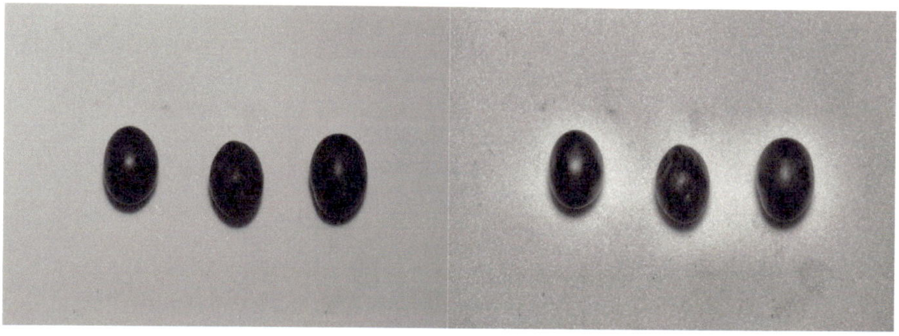

a) b)

Figure 12. Preprocessing of the image before applying the Hough transform. (a) Original image in grayscale. (b) Results of applying CLAHE equalization.

The sugar content of the grapes is commonly controlled by periodically measuring the content of soluble solids in the ripening berries with a refract meter. It works by measuring refracted light through grape juice in a prism. The sugar level is generally expressed in Brix grade, which represents grams of sugar per 100 grams of juice; the thicker juice shows a higher Brix degree on the scale.

The procedure followed by Murillo-Bracamontes et al. [3] in order to extract color parameters from berry image and to compare them against Brix degree (measured with a refract meter) as well as to their pH value was as follows: the acquired images were converted to grayscale and equalized using a contrast-limited adaptive histogram equation (CLAHE) to improve the contrast of grayscale images (see **Figure 12**) [71, 72]. A Hough transformation was then applied to locate the grapes (as shown in **Figure 13**). After completion of the transformation, three vectors corresponding to (x, y) and radius parameters of the berries were obtained; then, the

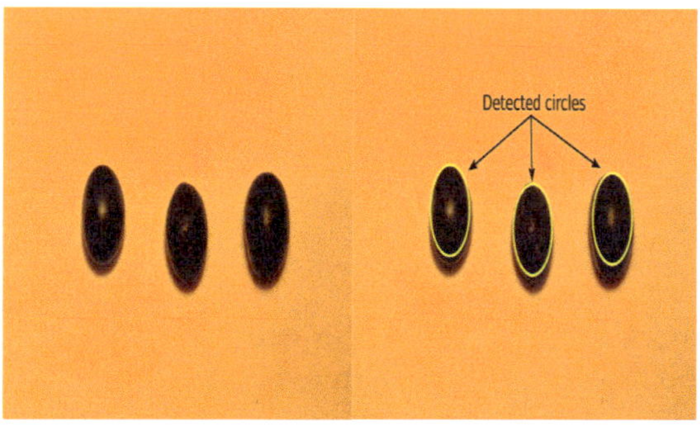

a) b)

Figure 13. Hough transform. (a) Original image in RGB color space. (b) Results of the detected circles (remarked in yellow).

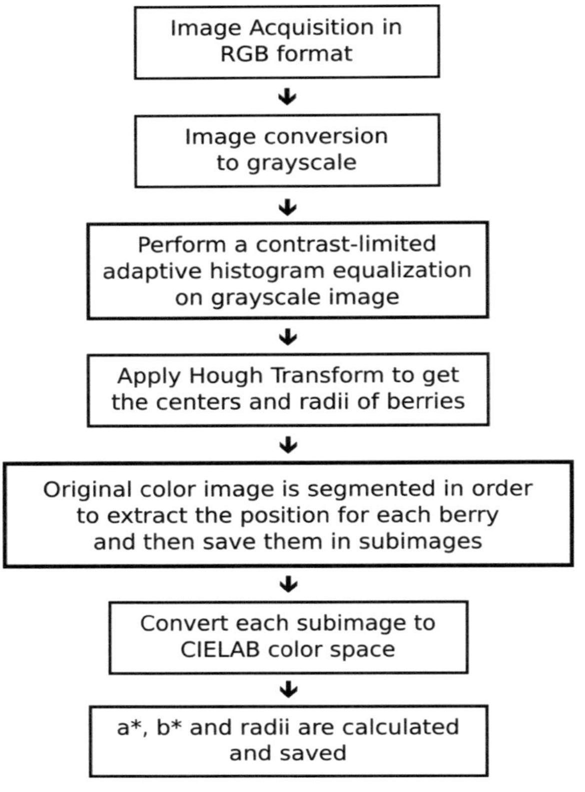

Figure 14. Algorithm applied to grape images.

individual berries detected were segmented from the background, stored in sub-pictures, and then converted to the CIELAB color space for analysis. **Figure 14** shows a flowchart of the steps performed for this analysis.

Analyzed data yield the same evolution described in **Figure 11**, which is a good indicator that the procedure provides promising values.

6. Conclusions

Computer vision systems are a viable option for handling agricultural products. The image processing and image analysis tools are well suited for application developments with automatic extraction of external and internal attributes of agro-products. One of the trends is the development of vision applications for in situ use in fields or greenhouses. The properties of being noninvasive and nondestructive make these systems an attractive option for application developments in different areas. However, agro-industry is particularly interesting due to the nature of their products and the standards of quality requirements.

http://dx.doi.org/10.5772/intechopen.71935

Acknowledgements

The authors would like to thank the National Council of Science and Technology (CONACYT) for providing the fundings to carry out this research, as well as to the people involved on this research at the Autonomous University of Baja California, Program for Strengthening Educational Quality (PFCE) and the Center for Scientific Research and Higher Education at Ensenada and the State University of Sonora.

Author details

Ernesto Martínez Sandoval[1], Miguel Enrique Martínez Rosas[1], Jesús Raúl Martínez Sandoval[2]*, Manuel Moises Miranda Velasco[1] and Humberto Cervantes De Ávila[1]

*Address all correspondence to: jesus.martinez@ues.mx

1 Autonomous University of Baja California, Ensenada, Mexico

2 State University of Sonora, Hermosillo, Mexico

References

[1] David H. Eye, Brain and Vision. W.H. Freeman; Scientific American Library, Henry Holt and Company, 1988. ISBN: 0716760096, 9780716760092

[2] Acharya T, Ray AK. Image Processing: Principles and Applications. John Wiley & Sons; 2005. pp. 425. ISBN: 0471745782, 9780471745785

[3] Murillo-Bracamontes EA, Martinez-Rosas ME, Miranda-Velasco MM, Martinez-Reyes HL, Martinez-Sandoval JR, Cervantes-De-Avila H. Implementation of Hough transform for fruit image segmentation. Procedia Engineering. 2012;**35**:230-239

[4] Davies ER. Machine vision, third edition: Theory, algorithms, practicalities. Pattern Recognition Letters. 2004;**7010**:934

[5] Timmermans AJM. Computer vision system for on-line sorting of pot plants based on learning techniques.International Society for Horticultural Science. 1998;**421**:91-98. DOI: 10.17660/ActaHortic.1998.421.8

[6] Sonka M, Hlavac V, Boyle R. Image Processing, Analysis and Machine Vision, First Edit. Springer-Science Business Media, B.V. 1993. pp. 555. ISBN: 978-0-412-45570-4, 978-1-4899-3216-7

[7] Jha SN. Nondestructive Evaluation of Food Quality. Springer US: India; 2010

[8] Bechar A. Agricultural robots for field operations: Concepts and components. Biosystems Engineering. 2016;**149**:94-111

[9] Tezmol A, Sari-Sarraf H, Mitra S, Long R, Gururajan A. Customized hough transform for robust segmentation of cervical vertebrae from X-ray images. In: Fifth IEEE Southwest Symposium on Image Analysis and Interpretation. 2002

[10] van den Broek WHAM, Noordam JC, Pauli A. Multivariate Imaging for Automated Process Control in the Agro Industry. IFAC Proceedings Volumes. 2000;**33**(19):303-307

[11] Marzotto R, Zoratti P, Bagni D, Colombari A, Murino V. A real-time versatile roadway path extraction and tracking on an FPGA platform. Computer Vision and Image Understanding. 2010;**114**(11):1164-1179

[12] Sentouh C, Popieul JC. Human-machine interaction in automated vehicles: The ABV project. In: Proceedings of the 19th World Congress the International Federation of Automatic Control Cape Town. Vol. 47, no. 3. 2014. pp. 6344-6349

[13] Davidson VJ, Ryks J, Chu T. Fuzzy models to predict consumer ratings for biscuits based on digital image features. IEEE Transactions on Fuzzy Systems. 2001;**9**(1):62-67

[14] Li J, Tan J, Shatadal P. Classification of tough and tender beef by image texture analysis. Meat Science. 2001;**57**(4):341-346

[15] Tan FJ, Morgan MT, Ludast LI, Forrest JC, Gerrard DE. Assessment of fresh pork color with color machine vision. Journal of Animal Science. 2000;**78**(12):3078-3085

[16] Davenel A, Seigneurin F, Collewet G, Rémignon H. Estimation of poultry breastmeat yield: Magnetic resonance imaging as a tool to improve the positioning of ultrasonic scanners. Meat Science. 2000;**56**:153-158

[17] Mery D et al. Automated fish bone detection using X-ray imaging. Journal of Food Engineering. 2011;**105**(3):485-492

[18] Costa C, Loy A, Cataudella S, Davis D, Scardi M. Extracting fish size using dual underwater cameras. Aquacultural Engineering. 2006;**35**:218-227

[19] Storbeck F, Daan B. Fish species recognition using computer vision and a neural network. Fisheries Research. Apr. 2001;**51**(1):11-15

[20] Elmasry G, Cubero S, Moltó E, Blasco J. In-line sorting of irregular potatoes by using automated computer-based machine vision system. Journal of Food Engineering. Sep. 2012;**112**(1-2):60-68

[21] Brosnan T, Sun D. Improving quality inspection of food products by computer vision–—A review. Journal of Food Engineering. 2004;**61**(1):3-16

[22] Kramer A. Evaluation of quality of fruits and vegetables. In: Food Quality. Irving GW, Hoover SR, editors. American Association for the Advancement of Science, Washington, DC; 1965. pp. 9-18

[23] Barrett DM, Beaulieu JC, Shewfelt R. Color, flavor, texture, and nutritional quality of fresh-cut fruits and vegetables: Desirable levels, instrumental and sensory measurement, and the effects of processing. Critical Reviews in Food Science and Nutrition. May 2010;**50**(5):369-389

[24] Huang C, Yang W, Duan L, Jiang N, Chen G, Xiong L. Rice panicle length measuring system based on dual-camera imaging. Computers and Electronics in Agriculture. 2013; **98**:158-165

[25] Huang M, Wan X, Zhang M, Zhu Q. Detection of insect-damaged vegetable soybeans using hyperspectral transmittance image. Journal of Food Engineering. 2013;**116**(1): 45-49

[26] Nguyen TT, Vandevoorde K, Wouters N, Kayacan E, De Baerdemaeker JG, Saeys W. Robotic agriculture detection of red and bicoloured apples on tree with an RGB-D camera. Biosystems Engineering. 2016;**146**:33-44

[27] Andújar D, Ribeiro A, Fernández-quintanilla C, Dorado J. Using depth cameras to extract structural parameters to assess the growth state and yield of cauliflower crops. Computers and Electronics in Agriculture. 2016;**122**:67-73

[28] Cho B et al. Detection of cuticle defects on cherry tomatoes using hyperspectral fluorescence imagery. Postharvest Biology and Technology. 2013;**76**:40-49

[29] Alirezaei M, Zare D, Nassiri SM. Application of computer vision for determining viscoelastic characteristics of date fruits. Journal of Food Engineering. 2013;**118**(3):326-332

[30] Baigvand M, Banakar A, Minaei S, Khodaei J, Behroozi-Khazaei N. Machine vision system for grading of dried figs. Computers and Electronics in Agriculture. 2015; **119**:158-165

[31] Ma J, Du K, Zhang L, Zheng F, Chu J, Sun Z. A segmentation method for greenhouse vegetable foliar disease spots images using color information and region growing. Computers and Electronics in Agriculture. 2017;**142**:110-117

[32] Lu J, Sang N. Detecting citrus fruits and occlusion recovery under natural illumination conditions. Computers and Electronics in Agriculture. 2015;**110**:121-130

[33] Kurtulmuş F, Ünal H. Discriminating rapeseed varieties using computer vision and machine learning. Expert Systems with Applications. 2015;**42**(4):1880-1891

[34] Annadurai S. Fundamentals of Digital Image Processing. Pearson, Pearson Education India; 2007

[35] Sun DW. Computer Vision Technology for Food Evaluation. Academic Press, Elsevier; 2007. ISBN: 0123736420, 9780123736420

[36] Christophe C, Marchand E. Colorimetry-based visual servoing. IEEE/RSJ Int. Conf. on Intelligent Robots and Systems, Oct 2009, St Louis, United States. 2009. pp. 5438-5443

[37] Sarkar NR. Machine Vision for Quality Control in the Food Industry: In Instrumental Methods for Quality Assurance in Foods. Food Science and Technology. Taylor & Francis; 1991. ISBN: 082478278X, 9780824782788

[38] Gunasekaran S. Computer vision technology for food quality assurance. Trends in Food Science and Technology. 1996;**7**(8):245-256

[39] Jhawar J. Orange sorting by applying pattern recognition on colour image. Procedia Computer Science. 2016;**78**(December):691-697

[40] Pearson TC, Slaughter DC. Machine vision detection of early split pistachio nuts. Transactions of the ASAE. 1996;**39**(3):1203-1207

[41] Steinmetz V, Roger JM, Moltó E, Blasco J. On-line fusion of colour camera and spectrophotometer for sugar content prediction of apples. Journal of Agricultural Engineering Research. 1999;**73**(2):207-216

[42] Chherawala Y, Lepage R, Doyon G. Food grading/sorting based on color appearance trough machine vision: The case of fresh cranberries. In: 2006 2nd International Conference on Information & Communication Technologies. vol. 1. 2006. pp. 540-545

[43] Feng G, Qixin C. Study on color image processing based intelligent fruit sorting system. In: Fifth World Congr. Intell. Control Autom. (IEEE Cat. No.04EX788). Vol. 6. 2004. pp. 4802-4805

[44] Garrido-Novell C, Pérez-Marin D, Amigo JM, Fernández-Novales J, Guerrero JE, Garrido-Varo A. Grading and color evolution of apples using RGB and hyperspectral imaging vision cameras. Journal of Food Engineering. 2012;**113**(2):281-288

[45] Abdullah MZ, Fathinul-Syahir AS, Mohd-Azemi BMN. Automated inspection system for colour and shape grading of starfruit (*Averrhoa carambola* L.) using machine vision sensor. Transactions of the Institute of Measurement and Control. 2005;**27**(2):65-87

[46] Cho JS, Lee HJ, Park JH, Sung JH, Choi JY, Moon KD. Image analysis to evaluate the browning degree of banana (Musa spp.) peel. Food Chemistry. 2016;**194**:1028-1033

[47] Esehaghbeygi A, Ardforoushan M, Monajemi SAH, Masoumi AA. Digital image processing for quality ranking of saffron peach. International Agrophysics. 2010;**24**:115-120

[48] Noordam JC, Otten GW, Timmermans TJM, van Zwol BH. High-speed potato grading and quality inspection based on a color vision system. Proceedings of SPIE. 2000;**3966**: 206-217

[49] Barnes M, Duckett T, Cielniak G, Stroud G, Harper G. Visual detection of blemishes in potatoes using minimalist boosted classifiers. Journal of Food Engineering. 2010;**98**(3): 339-346

[50] Carlos L, Antonio L, Sanches J, Maria DAL, Fabbro I. Image processing techniques for lemons and tomatoes classification. Bragantia, Campinas. 2008;**67**(3):785-789

[51] Moreda GP, Ortiz-Cañavate J, García-Ramos FJ, Ruiz-Altisent M. Non-destructive technologies for fruit and vegetable size determination—A review. Journal of Food Engineering. May 2009;**92**(2):119-136

[52] Brosnan T, Sun D. Inspection and grading of agricultural and food products by computer vision systems—A review. Computers and Electronics in Agriculture. 2002;**36**:193-213

[53] Carosio MGA, Bernardes DF, Andrade FD, Moraes TB, Tosin G, Colnago LA. Measuring thermal properties of oilseeds using time domain nuclear magnetic resonance spectroscopy. Journal of Food Engineering. 2016;**173**:143-149

[54] Soares Valeria LM, Alves Filho EG, Mara A. Silva L, Henrique Novotny E, Marques Canuto K, Jair Wurlitzer N, Narain N, Sousa de Brito E. Tracking thermal degradation on passion fruit juice through Nuclear Magnetic Resonance and chemo-metrics, Food Chemistry. 2016;**219**:1-6

[55] Zhang Y, Wu L. Classification of fruits using computer vision and a multiclass support vector machine. Sensors. 2012;**12**(9):12489-12505

[56] Nagata M, Tallada JG, Kobayashi T. Bruise detection using NIR hyperspectral imaging for strawberry (FragariaXananassa Duch). Environmental Control in Biology. 2006; **44**(2):133-142

[57] Unay D, Gosselin B, Kleynen O, Leemans V, Destain MF, Debeir O. Automatic grading of bi-colored apples by multispectral machine vision. Computers and Electronics in Agriculture. 2011;**75**(1):204-212

[58] ElMasry G, Wang N, Vigneault C. Detecting chilling injury in red delicious apple using hyperspectral imaging and neural networks. Postharvest Biology and Technology. 2009;**52**(1):1-8

[59] Gómez-Sanchis J, Martín-Guerrero JD, Soria-Olivas E, Martínez-Sober M, Magdalena-Benedito R, Blasco J. Detecting rottenness caused by Penicillium genus fungi in citrus fruits using machine learning techniques. Expert Systems with Applications. 2012;**39**(1): 780-785

[60] Khoje S, Bodhe S. Performance comparison of Fourier transform and its derivatives as shape descriptors for mango grading. International Journal of Computers and Applications. 2012;**53**(3):17-22

[61] Taghizadeh M, Gowen AA, O'Donnell CP. The potential of visible-near infrared hyperspectral imaging to discriminate between casing soil, enzymatic browning and undamaged tissue on mushroom (*Agaricus bisporus*) surfaces. Computers and Electronics in Agriculture. 2011;**77**(1):74-80

[62] Larhmam M, Mahmoudi S, Benjelloun M. Semi-automatic detection of cervical vertebrae in X-ray images using generalized Hough transform. In: Proceedings of the 3rd International Conference on Image Processing Theory, Tools and Applications.IPTA'12. 2012. pp. 396-401

[63] Donis-González IR, Guyer DE, Pease A, Barthel F. Internal characterisation of fresh agricultural products using traditional and ultrafast electron beam X-ray computed tomography imaging. Biosystems Engineering. 2013;**7**:104-113

[64] Chuang C et al. Automatic X-ray quarantine scanner and pest infestation detector for agricultural products. Computers and Electronics in Agriculture. 2011;**77**(1):41-59

[65] Chamelat R, Rosso E, Choksuriwong A, Rosenberger C, Laurent H, Bro P. Grape detection by image processing. In: IECON 2006—32nd Annual Conference on IEEE Industrial Electronics. 2006. pp. 3-8

[66] Martinez-Sandoval JR, Murillo-Bracamontes EA, Martinez-Rosas ME, Miranda-Velasco MM, Cervantes De Avila H. Image processing ap-plied in agriculture. In: Raul AS, Edwards Block A, editors. Embedded Systems and Wireless Technology. 1st ed. Boca Raton: CRC Press; 2013. pp. 201-226. ISBN: 9781466565654

[67] Rabatel G, Guizard C. Grape berry calibration by computer vision using elliptical model fitting. In: ECPA 2007, 6th Eur. Conf. Precis. Agric. 2007

[68] Nuske S, Achar S, Bates T, Narasimhan S, Singh S. Yield estimation in vineyards by visual grape detection. IEEE International Conference on Intelligent Robots and System. 2011:2352-2358

[69] Hellman E. How to judge grape ripeness before harvest. In: Southwest Regional Vine and Wine Conference. 2004

[70] Dami I et al. Midwest Grape Production Guide. Columbus, OH: Ohio State University Extension; 2005

[71] Sonia Goyal S. Region based contrast limited adaptive HE with additive gradient for contrast enhancement of medical images (MRI). International Journal Computer Science Engineering. 2011;**1**(4):154-157

[72] Garg R, Mittal B, Garg S. Histogram equalization techniques for image enhancement. International Journal of Electronics Communication and Computer Technology. 2011;**2**(1):107-111

Chapter 8

Automation of Integrated System for Grain Beverages Processing

Gana Ibrahim Mohammed, Ephraim Micheal and Agidi Gbabo

Additional information is available at the end of the chapter

http://dx.doi.org/10.5772/intechopen.73248

Abstract

The research work focus on design and construction of automatic system for integrated plant for grain beverages processing. A grain beverage processing plant is a complex system that integrates several operations (blending of soaked grains, mixing the slurry, extracting the aqueous liquid and discharging of the paste out of the machine) together and finished in one go. Incorporating an automatic system into the integrated system simplify its mode of operation. Essential design consideration, analysis and calculations were carried out in order to determine and select materials of appropriate strength and sizes for various part of the automatic system. The major parts of the automatic system includes power supply unit, transformer, filter capacitor, voltage regulator, power indicator, pre-set buttons, time controller, eprom, display unit, controllers, limiting sensor, solenoid valve and electro-mechanical switch. The system was designed to have two controllers, one interfaced with the button network and the other organized the operational time (blending, sieving and paste expelling) in minute. Results of the testing revealed that the highest machine output of 90.24 L/h was obtained from speed of 1650 rpm using the integrated machine with automated system, low value of output of 52.64 L/h was obtained from the same speed using the integrated machine with semi-automatic system. The least machine output of 32.59 L/h was obtained from the same speed using the integrated machine without the automated system. The machine output was found to be influenced by both the automatic system and machine speed. The automatic system allows efficient work flow, reduces human labor, ensure safety and hygiene product production by eliminating human interference. Also it increased the machine output by 67%, reduce operational time by 65% and completely eliminating human interference with the product.

Keywords: automated, beverages, grain, integrated, system

1. Introduction

Automation in farm machineries and processing is a very important factor considered to reduce losses, achieve faster and better ways of food processing so as to meet the increasing demands of consumers. Food processing is an important operation that contributes immensely in economic development of the states as it is vital in ensuring food availability and security all over the globe. According to Gana [1] the production process of most agricultural food materials is multi operational process, comprises of many different unit operations requiring separate equipment.

The present trend in agricultural food manufacturing and processing is focused on automation of integrated system that combines many batch operations into single manufacturing system. The design provides on-line and continues control capacity. According to Gunasekaran [2] most of the automation systems carried out in food industries are isolated, batch-type operations targeting a specific task. Steve [3], reported that food processing and manufacturing operations in small and median enterprises (SME) are basically carried out manually unlike in larger industries where automations were achieved either with the aid of robots or combination of simpler electromechanical devices [3].

Food Processing has been defined by Gana [1] as the alteration of raw food materials into consumable state or the later into other forms. According to Rachel [4] food processing involves the use of clean raw materials either from crops or animal product to produce good-looking and profitable products and animal feed. It also helps in extending the shelf-life of these products. Food processing removes toxic materials from food, enhanced preservation, marketing, increased food concentration and availability of several foods which are beneficial to the consumers. The author stated that health standard of certain group of people with specific health problems such as diabetes and allergies can be improved through modern food processing. Also additional nutrients can be added to certain class of food that lacks such nutrients. It sometimes involves mechanical processes that employ the use of mixing and grinding equipment and machines in the production line. The author also stated that in food processing industry, the food performance parameters are vital element necessary in the design process. Some of these parameters include: hygiene, energy efficiency, labor used [4].

1.1. Importance of automation in food processing

According to Jijo and Ramesh Kumar [5] some of the advantages of automation in food processing include improved productivity in processing line by allowing efficient schedule of work flow and labor utilization. Also it ensures high quality products consistently thereby encourage customer loyalty and this result to expanding market share. In addition one of the major advantages of automation is ensuring food hygiene and safety by eliminating human interference with food product.

For successful automation of agricultural food manufacturing and processing an integration of the manufacturing process must be carried out with view of transforming the operations into single manufacturing design.

1.2. Unit operations involved in beverages production from grains

The production of beverages from grains involved series of unit operations. The major unit operations include steeping, size reduction (milling), mixing, sieving, filtration, paste discharge, sedimentation, boiling, storage and packaging [1].

1.3. The present status of grain beverages production

The available machines are made from mild steel materials and with frequent contact with water, rusting easily take place. This can easily lead to contamination of the food materials thus decreasing quality of the final food product. Also the production process involve different steps using various machine and equipments that make the production procedure tedious, time consuming and products are predisposed to contamination as results of human interaction and exposure to the environment. In additional there is possibility of contamination of beverages as result of wearing of component parts of milling plate with time of use. Furthermore, unlike milling machine the sieving machine is not easily accessible by the local processors. It is expensive and its mode of operation is complex [6]. Therefore there is need to develop an automated machine plant capable of integrating the various unit operations such as blending soaked grains, mixing of the slurry, extracting the aqueous liquid and expelling the paste out of the machine into single operation [1, 7].

Hence, the design and construction of an automatic system for grain beverages processing plant capable of blending soaked grains, mixing of the slurry, extracting the aqueous liquid and expelling the paste out of the machine in one compartment became necessary in order to improve on the functionality of the already existing grain drinks processing machine.

2. Materials and methods

2.1. Materials selection

A Grain beverages processing plant developed by Gana et al. [7] was used in this study. Grain processing plant is complex machines that can blend soaked grains, mixed the slurry with water, sieved, extract the aqueous liquid and expelled the paste out of the machine. The machine is shown in **Figure 1**.

2.2. Integrated system description

The system is made up of the following components; outer casing made up of stainless sheet and a liquid out let valve was fitted to its bottom side in order to allow out flow of extracted aqueous liquid from the tank. A paste outlet chute was also fitted at the bottom of the casing for discharging of expelled paste as shown in **Figure 2**. Conical centrifugal basket with lower impervious part that allowed thorough washing of the milk from the paste and upper part which is smooth with perforated openings in order to allow fluid drainage and also to prevent paste losses as shown in **Figure 2**. A conical screen was also fitted inside the conical

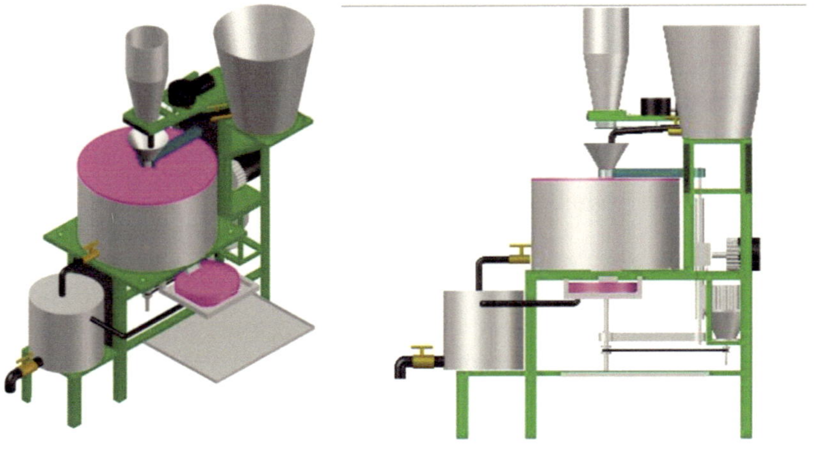

Figure 1. AutoCAD drawing of the machine.

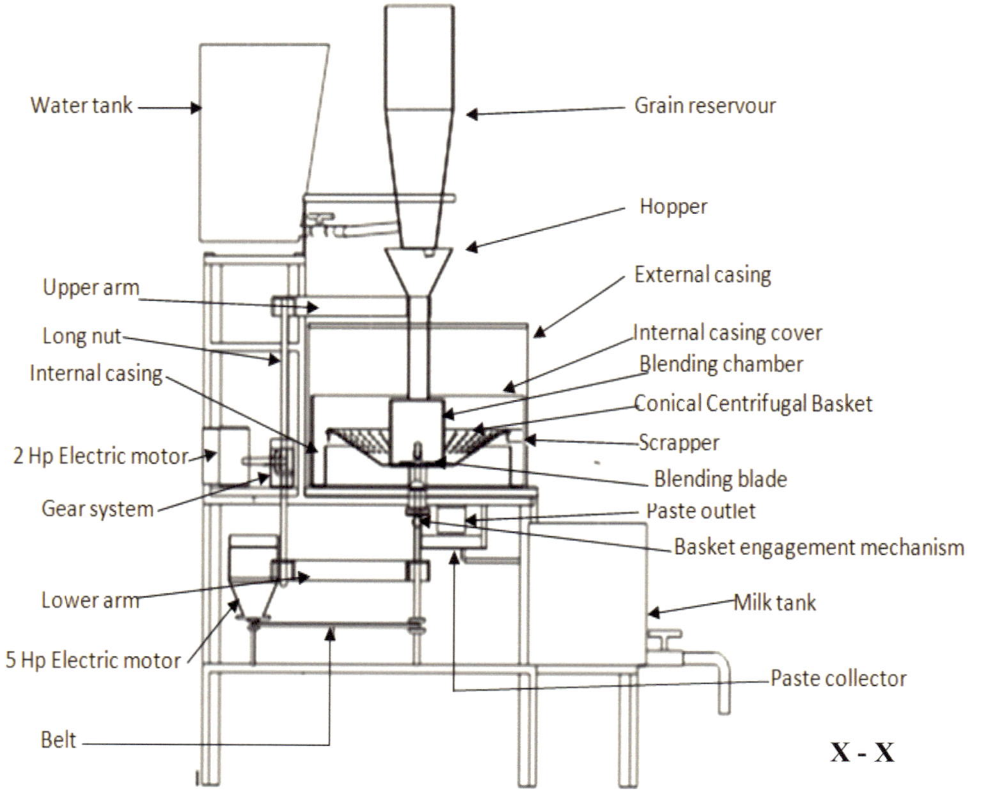

Figure 2. Cross section of the integrated system for grain beverages processing.

centrifugal basket in order to allow fluid drainage and prevent paste losses. Scrapper was attached to the conical basket at the top from outside. It scraped, conveyed and discharged the expelled paste from the basket through the discharge outlet as shown in **Figure 2**. The Internal casing was attached to the internal wall of the outer casing. It is cylindrical in shape with its upper side opened in order to allow collection of expelled pastes. The hopper serves as the reservoir where the grains are fed to the machine. It was made up of stainless sheet, and of conical shape as shown in **Figure 2**. The delivery pipe is cylindrical in shape and conveyed the grains directly to the blending chamber; it is shown in **Figure 2**. Blending Chamber is where the blending operation takes placed. It was designed to prevent the materials from spilling and moving out of the blending chamber until after the blending operation is completed, it is shown in **Figure 2**. Blending Blade this is attached to the shaft inside the conical basket. The gear box this controlled the movement of the upper and lower arms. Therefore is responsible for opening and closing of the blending chamber, engagement and disengagement of the basket from rotation [1, 7].

2.3. Design plan of the automation

To automate the system, there is need to have blending and sieving time as both operations cannot occur at the same time. To ensure flexibility, the timing system designed must be programmable so as to ensure the best blending by adjusting the blending time. During blending, the blending chamber has to be closed in order to ensure efficient blending and also the centrifugal basket has to be disengaged from rotation with the central shaft. The flow of the process is shown in **Figure 3** while the block diagram that illustrates the hardware design unit by unit and their interconnection is shown in **Figure 4**.

2.4. Components of automatic system

The Automatic system is a combination of electronic, electrical and mechanical parts. Some of the parts include the following;

Power unit; this unit was made up of the following components: transformer, rectifier, filtering capacitor, voltage regulator and power indicator [8].

Transformer; the transformer used for this purpose was a 15 V 3 A transformer due to its availability in the local store. It steps down the current from 220 to 15 V A.C. A circuit representation of this component is shown in **Figure 5**. A bridge rectifier was employed for this design in order to ensure a smother conversion of the current from A.C. to D.C. However it is important to note that despite the conversion, some element of A.C was still observed with the D.C. power supply [9, 10].

Capacitor; the filter capacitor helps to ensure the complete conversion of A.C. power supply to D.C. power supply. The electrolytic component has to be carefully selected based on filter capacity and dielectric voltage. Usually, the dielectric voltage which is often written on the body of the capacitor must be greater than the voltage supply to the component when connected to avoid damage, it is shown in **Figure 5** [11].

Voltage regulator; this help to achieved steady voltage supply to drive the whole control system. This parameter is often times influenced by the demand of the circuit. In this design, 5 V is demanded by the controllers which by manufacturer's instruction can be powered with 3 to 5 V for effective workability. For this reason 7805 voltage regulator was used to power the system as shown in **Figure 5** [12].

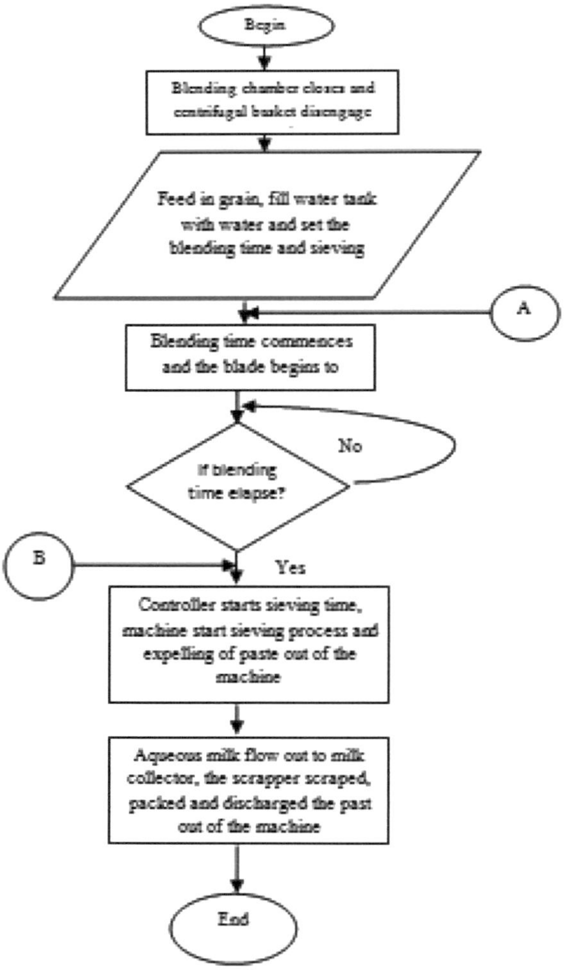

Figure 3. The flow process of the system.

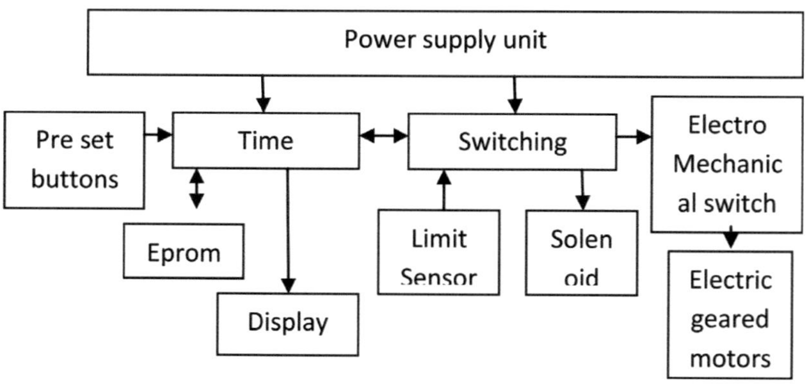

Figure 4. Block diagram of the system.

http://dx.doi.org/10.5772/intechopen.73248

Power indicator; the purpose of this part of the circuit is to give a visual evidence of the state of live of the power supply. This is achieved via the use a resistor connected in series with a light emitting diode (**Figure 5**) as reported by Thakur and Sharma [13].

Pre-set buttons; this unit helps to preset the blending time and paste expelling time. To achieve this, the switch is connected in series with a resistor so as to achieve difference in logic at different state of the switch. When the switch was opened it is logic HIGH (4.3 V) and when close is logic LOW (0 V), it is shown in **Figure 5** [14].

Time Controller; the time controller (Atmel 89C52) used is compactable with MCS-51. The 8 K byte controller in terms of in-system reprogrammable flash memory can be reprogrammed 1000 times, when the clocking frequency is between 0 Hz and 24 MHz. The device which has 32 bits input-output pins, three 16 bit counter/timer, eight interrupt sources, programmable serial channel is embedded with 252 × 8 bit internal RAM. It is however connected as seen in **Figure 5** according to manufacturers' instruction and programmed using Kiel version IDE for effective functioning. Pin 31 is connected to the V_{CC} (5 V) so as to ensure that the controller fetches instructions from its internal program memory. The reset pin (pin 9) is connected to the mid-point of a resistor capacitor series network. This ensures delay while booting so that the controller fully booths up before performing the instructions. Pins 18 and pin 19 is connected to crystal oscillator of 12 MHz. It is shown in **Figure 5**.

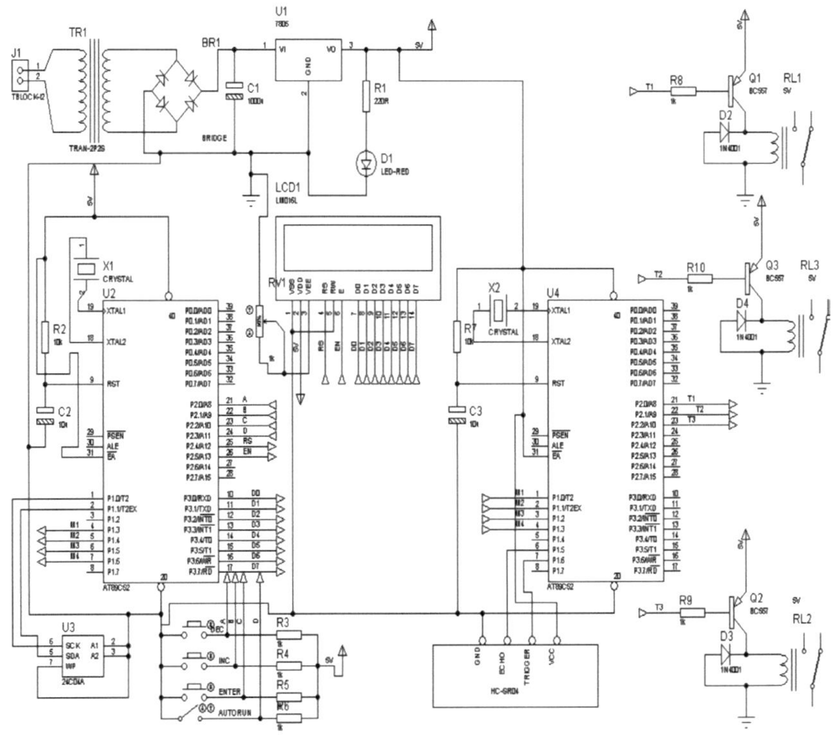

Figure 5. Complete circuit diagram of the system.

Eprom; the Eprom (24C08) used in this project stored the preset data such as the blending time, and paste expelling time. The device is connected to the controller in such a way that it communicates using I^2C. The essence of using the Eprom is to prevent loss of data due to power failure.

Display unit; the display unit used in this work is an alpha numeric 16 by 2 LCD display, it serves as medium of interaction with the operator. The 16 pin device (**Figure 5**) could be interfaced with the controller using 1 byte (using DB0 to DB7) or 1 nibble (4 bits using DB4 to DB7). However to aid a fast display, the 1 byte medium of communication is used. Pin 1 and pin 2 is the power pin of the display connected to ground and V_{CC} respectively. The contrast of the display is adjustable via the use of a variable resistor connected between pin 3 and the ground. According to the data sheet, the logic on pin 4 which is the register select pin and determines if the LCD will display the data sent to it or use it as instructions. When this pin is connected to a pin on the controller and it becomes logic zero (0 V), the LCD takes the data sent to it as an instruction. But if it is logic 1, it takes the data as information to be displayed. Pin 5 which is the read or write pin is connected to ground since the intention is to write not read [13].

Switching Controller; this unit interfaced with the relays and responsible for the relay (mechanical switches) actions which are used to control the electric motors. Atmel 89C52 used in this unit is also interfaced with the time controller unit so as to know what operation or motor to control at a particular time. The basic connection of this unit is like that of the time controller unit. It is shown in **Figure 6**.

Limiting sensor; the limiting sensor (ultrasonic module HC-SR04) helps to determine the maximum distance the arms will be move upward in order to open the blending chamber and to engage the basket in rotation with the shaft and also the distance the arms will move downward in order to close the blending chamber and to disengage the basket in rotation with the shaft. This is interfaced with the switching controller so that whenever the command is given to move the arms either upward or downward, it will stop motion when the exact distance is

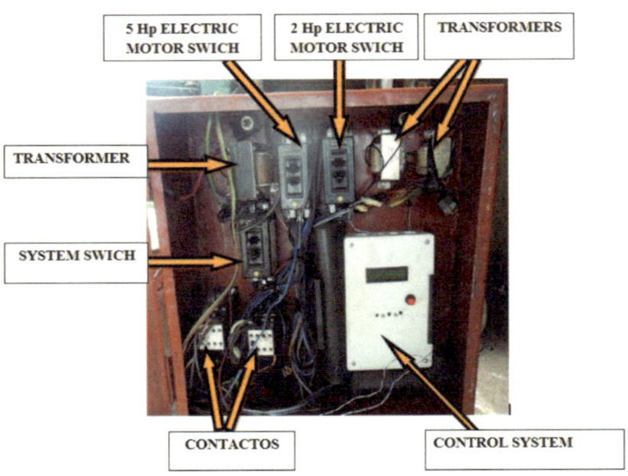

Figure 6. Component part of the automated system.

http://dx.doi.org/10.5772/intechopen.73248

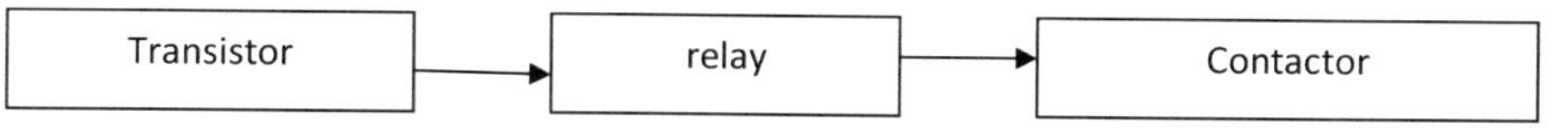

Figure 7. Block diagram of the electromagnetic switch unit.

moved. The four pin device receives a trigger pulse of 10 μs from the controller to transmit a burst of 40 KHz. If an obstacle is seen in line of sight, the ultrasonic burst is reflected at the echo pin giving a pulse width which ranges from 120 μs to 25 ms. However if no obstacle is detected, the echo pulse width remains at 38 ms. This echo pulse width is then read by the controller and the actual distance determined [15, 16].

Electro-Mechanical switch; this unit is called the actuator, it consist of driving relays interfaced to the switching controller via a transistor as shown in **Figure 7**. The reason for driving the contactors with a relay is because they (contactors) are A.C. driven so cannot be driven directly with the controller. For this project tow contactors were used to change the direction of AC motor that aids the opening and closing of the blending chamber and to control the spinning of the blade to blend and stop blending [14].

Electric motor; the two electric motors used in this design are A.C powered motors. The 2 Hp electric motor was used to close and open the blending chamber as while as for engagement and disengagement of conical basket in rotation with the central shaft. While the 5 Hp electric motor was used to operates the central shaft, blending blade and conical centrifugal basket.

2.5. System design analysis

2.5.1. The power supply circuit

A step-down transformer with turn's ratio of 16:1 was selected to transform the 240 V mains supply voltage to 15 V for the power supply. The 15 V ac was converted to dc voltage using a full wave rectifier circuit. The circuit was designed as reported by Agbetuyi and Orovwode [17], and is given as follows;

$$\gamma = \sqrt{V_{dc}^2 - V_{ac}^2} / V_{dc} \tag{1}$$

$$V_{dc} = V_m / 1.414 \tag{2}$$

$$V_{ac} = (2/\pi) \times V_m \tag{3}$$

$$V_m = V_{sp} - V_b \tag{4}$$

$$V_{sp} = 1.414 \times V_s \tag{5}$$

$$V_b = 2 \times V_{dd} \tag{6}$$

Where, γ is ripple factor for a full wave rectification process using a diode bridge (V), V_{dc} is rms value of output dc voltage of the diode bridge (V), V_{ac} is average value of the diode bridge output voltage (V), V_m is peak output dc voltage from the diode bridge (V), π is constant (3.142), V_{sp} is peak value of transformer secondary voltage (V), V_s is transformer secondary voltage (V), V_b is voltage drop across the diode bridge at any instant (V), V_{dd} is diode forward conduction voltage drop (V).

The ripple in the output voltage is directly proportional to the output current and is related to the filtering capacitance by the following equations as reported by Agbetuyi and Orovwode [17],

$$C = I/(2 \times f \times V_r) \tag{7}$$

$$V_r = \gamma \times V_{sp} \tag{8}$$

But $dV_{sp} = V_r$

$$f = t/2 \tag{9}$$

$$q = I \times t = C \times dV_{sp} \tag{10}$$

Where, C is capacitance value (μF), I is required output current from power supply circuit (A), t is time taken for filtering capacitor to discharge in compensation for the ripple in the dc output (s), f is frequency of the ac mains supply voltage (Hz), q is charge on filtering capacitor (A), dV_{sp} is peak value of transformer secondary voltage (V).

2.5.2. The sensing circuit

The sensor designed was made up of two conducting metal plate of 7 cm length and 5 mm width each placed on the machine frame above and below the upper arm of the engagement mechanism. The distance of separation of each of the plate from the arm was 6 cm. The plates were connected to the circuit.

2.5.3. The control circuit

A 555 controller was used as the brain of the control circuit due to its operational characteristics in the monostable mode. The timer was used to produce 11.01 V to energize the relay coil, consequently activating the 2 Hp motor circuit for 7 s, which is the time needed for the blending chamber cover to open for 25 mm when the upper arm is moving upward (expelling of paste from blending chamber) and also to close that space when the upper arm is moving downward (for blending operation to take place)

$$R_t = T_{EB}/1.1 \times C_t \tag{11}$$

R_t is resistance tying the discharge and threshold pins to V_{cc} (V), T_{EB} is the time needed for opening and closing of the blending chamber cover during expelling and blending operation respectively (s), C_t is capacitance tying the discharge and threshold pins to ground (μF).

2.6. Working procedure of integrated system

The integrated system was designed to be operated in three forms with; the developed automatic system, semi-automatic system and without the automatic system.

2.6.1. Operating the integrated system with automatic system

The controller U2 interfaced with the button network (enter, INC and DEC), Eprom (U3), LCD (LCD1) and the other controller U4 is used to organize the time for blending, sieving and paste expelling time in minute. Firstly, when the system is powered, it will give a welcome message before asking for the operational (blending and paste expelling) time setting. When all these settings are done, the autorun switch is closed to save the parameters into the Eprom and enable the system into normal operation (**Figure 8**). During the operation, after the grains have been fed into the machine, and the water tank filled, the controller U2 through P1.3 which is connected to P1.0 of U4 sends a control signal (Logic HIGH) to U4 for the period

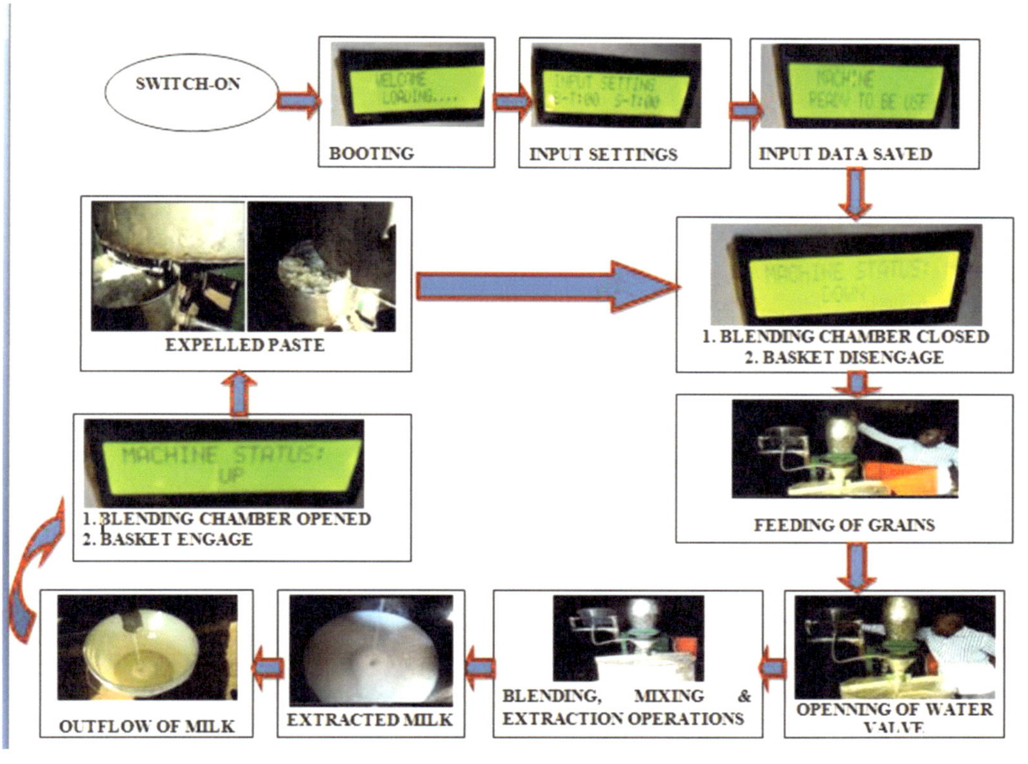

Figure 8. Mode of operation of the automated system.

of the blending time. This aids the energizing of relay (RL1) which links the contactor to AC power supply, powering the motor for the shaft to close the blending chamber. Three seconds later the line M2 which connects P1.4 of U2 to P1.1 of U4 becomes HIGH to start the blending. When the blending time is exhausted, M1 and M2 goes LOW and then the controller U2 makes M3 goes HIGH to start sieving and paste expelling.

2.6.2. Operating the integrated system with semi-automatic system

The system can also be operated in semi-automatic mode using the two contactors without the control system. This was achieved by switching on the first contactor which in turn activates the gear system to rotate in cloanti close wise direction, thereby moving the arms downward to close the blending chamber and to disengage the basket from rotation with the central shaft, for blending operation. After the blending operation is completed the first contactor is switched off and the second is switched on to activates the gear system this time to rotate in anti-clockwise direction. Thereby moving the arms upward to open the blending chamber and to engage the basket in rotation with the central shaft for sieving and paste expelling operations. The shortcomings of operating the machine in semi-automatic mode is that the operator has to monitor the upward and downward movement of the arms to ensure proper closing and opening of the blending chamber, as while as the engagement and disengagement of the basket. Also the blending and paste expelling time has to monitor. Unlike when the system is operated with the automatic system here the engagement and disengagement of the basket required that the machine most be stooped.

2.6.3. Operating the integrated system manually (without the automatic system)

In the mode of operation all the settings are carried out manually. Firstly, the upper arm that holds the hopper assembly is lower down until the blending chamber cover rested on the base of the basket. At the same time the lower arm is lower down until the basket is disengage from the shaft. The same setting is also required in opposite direction for sieving and paste expelling operations. The shortcomings of operating the machine manually is that all the settings were carried out manually which is tedious, time consuming and required expertise.

2.7. The constructed automatic system

The automatic system was designed, constructed and embedded to the integrated system as shown in **Figure 9**, while the automatic system and its accessories are shown in **Figure 6**.

2.8. Testing of the automatic system

The performance of the automatic system was evaluated in accordance with procedure reported by Gana et al. [7]. The soya bean (TGX 1954-IFXTGX 1835-10E) was purchased from Bida central market and the samples were cleaned and sorted to remove unwanted materials before soaking at room temperature of 27°C for the recommended time of 12 h [18] before processing using the automated integrated system. Three sets of experiments were carried out to investigate

Figure 9. Front and back view of the machine showing the automated system.

the system performance. In the first experiment the integrated system was operated without automation. In this case the opening and closing of the blending chamber covers as while as the engagement and disengagement of the conical centrifugal baskets in rotation with the central shaft were carried out manually. In the second experiment the integrated system was operated with semi-automatic system with aid of two contactors that control the opening and closing of the blending chamber covers as while as the engagement and disengagement of the conical centrifugal baskets in rotation with the central shaft. In the third experiment the integrated system was operated with the automatic system. In each of the experiment the effect of machine operational speeds (850, 1050, 1250, 1450 and 1650 rpm) on the machine output per hour was investigated. The experiments were carried out at the Agricultural and Bioenvironmental Engineering Department of Federal Polytechnic Bida, Nigeria.

2.9. Determination of performance parameters

The machine performance was determined based on machine output per hour.

2.9.1. Machine output

This is the quantity of aqueous liquid obtained after processing the grains. Procedure described in AOAC [19] and reported by Adebayo et al. [20] was used to evaporate water content of the aqueous milk in order to obtain soya milk with total solid content of 7% as recommended by SFAA [21]. It was measured in liters per hour (L/h) as reported by Gbabo et al. [22].

$$M_{OP} = N_{BH} \times \left(\mathrm{W}_{\mathrm{Ac}}/\mathrm{W}_{\mathrm{g}} + \mathrm{W}_{\mathrm{w}}\right) \quad (12)$$

where, M_{OP} is the machine output (L/h), N_{BH} is number of batches in 1 h, W_{Ac} is weight of the aqueous milk with 7% total solid content (Lts), W_{g} is weight of soaked soya beans processed (kg), W_{w} is weight of water used (kg).

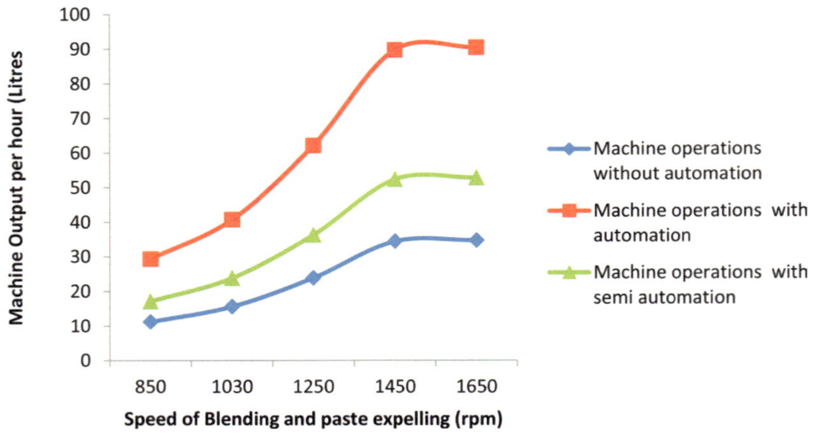

Figure 10. Results of effect of speed and automation system on machine output.

2.10. Experiment results

In all the three testing five levels of speeds of 850, 1050, 1250, 1450 and 1650 rpm were used based on an earlier findings by Gbabo et al. [22] and Gana et al. [7] to determine the effects of speed of machine operation as while as the effects of the automation on the machine output per hour. Each of the experiment was replicated three times using Eq. (12) and the results obtained are presented in **Figure 10**.

3. Results and discussion

3.1. Results

The automated system was designed, constructed and the result of the performance testing is presented in **Figure 10**. The highest machine output of 90.24 L/h was obtained from speed of 1650 rpm using the integrated machine with automated system, low value machine output of 52.64 L/h was obtained from speed of 1650 rpm using the integrated machine with semi-automatic system. The least machine output of 32.59 L/h was obtained from the same speed of 1650 rpm using the integrated machine without the automated system.

3.2. Discussion

3.2.1. Effects of automation on the integrated system (machine) output

The machine output was influenced by its operational speed and as well as by the automatic system. From **Figure 10**, high values of machine output of 29.28 L/h from speed of 850 rpm, 40.68 L/h from speed of 1050 rpm, 62.04 L/h from speed of 1250 rpm, 89.52 L/h from 1450 rpm and 90.24 L/h were obtained when the machine was operated with the automatic system. These values were greater more than values of 17.08, 23.73, 36.19, 52.22, and 52.64 L/h obtained from corresponding speed when the machine was operated with semi-automatic system. Least values of machine out of 11.22, 15.59, 23.78, 34.32, and 34.57 L/h were obtained from corresponding speed when the machine was operated without the control system. The higher values obtained from the former could be as result of elimination of manual settings of the upper and lower arms, closing and opening of the blending chamber cover for blending and paste expelling operations after the milk extraction respectively [1]. The time required for disengaging and engaging the centrifugal basket in rotation with the central shaft for blending and paste expelling operations respectively were eliminated. This indicated that the automatic system therefore reduce operational time by allowing efficient work flow and reduce human labor required. This is in line with the report of Jijo and Ramesh Kumar [5] were automation in food processing was found to improved productivity in processing line by allowing efficient schedule of work flow and labor utilization. The automation of the integrated system also ensure safety and hygiene by eliminating human interaction with the product thereby reduce possibility of contamination of product by human interaction and other factors such as dust, insects among others. This agreed with the report of Nayik et al. [23], where hygiene and cleanness of produced product are among the benefits of automation of food processing plant.

The machine output in both cases was observed to increase with increase in speed of operation. The total time required to process one batch was found to be 5 and 13 min when the machine was automated and non-automated respectively. The machine carried out 12 batches of operation in 1 h when automated and 4 batches in 1 h without automation. Therefore, it increased the machine output by 67%, reduce operational time by 65% and completely eliminating human interference with the product.

3.2.2. *Effects of integrated system (machine) operational speed on machine output*

From **Figure 10**, in all the three experiments the machine output increased significantly with increase in the machine operational speed. This could be owing to increase in impact force, cutting and shearing actions of the blade with increased in machine speed. Jayesh [24] and Gana [1] had reported that machine operational speed was found to be a key factor to segregation of solid materials. Where higher machine operational speed resulted to higher segregation of materials, while lower machine operational speed resulted to low segregation of materials. It was observed that there is significant ($p \leq 0.05$) differences between the machine output obtained from speed of 850 and 1650 rpm. But no significant ($p \leq 0.05$) differences were observed from machine output obtained from speed of 1450 and 1650 rpm. This could be as results of finer particle produced by speed of 1650 rpm, which clumped together and formed larger particles that clogged the sieve holes. As result of this some aqueous liquid were discharged out together with the paste. This agreed with the result of an earlier study by Douglas [25] where high speed of blending was found to produce finer particles in slurry. This particles clogged together and blocked the sieve holes, thus prevent materials from passing through the holes.

4. Conclusions

A control system for automation of an integrated system for grain beverages processing has been developed and tested. The controller U2 of the system was designed to interfaced with the button network (enter, INC and DEC), Eprom (U3), LCD (LCD1). The second controller U4 was designed to organize the time for blending, sieving and water dispensing time in minute. Test results of the automatic system on the machine revealed that the highest machine output of 90.24 L/h was obtained from speed of 1650 rpm using the integrated machine with automatic system, value of 52.64 L/h was obtained from the same speed when the integrated machine was used with semi-automatic system and low value of machine output of 34.59 L/h was obtained from the same speed when the integrated machine was used without the automatic system. With the automatic system the manual settings and time required for the settings of the machine parts were eliminated. The automated system allows efficient work flow, reduces human labor, it ensures safety and hygiene product. It increased the machine output by 67%, reduce operational time by 65% and completely eliminating human interference with the product. Also the machine output was found to increase with increase in machine operational speed.

Author details

Gana Ibrahim Mohammed[1]*, Ephraim Micheal[2] and Agidi Gbabo[3]

*Address all correspondence to: ganaibro74@yahoo.com

1 Department of Agricultural & Bio-environmental Engineering, Federal Polytechnic Bida, Niger State, Nigeria

2 Department of Research and Development, Scientific Equipment Development Institute Minna, Niger State, Nigeria

3 Department of Agricultural & Bioresources Engineering, Federal University of Technology Minna, Niger State, Nigeria

References

[1] Gana IM. Development and optimisation of functional parameters of an automated grain drinks processing machine [Unpublished PhD. Engineering Thesis]. Minna: Federal University of Technology; 2016

[2] Gunasekaran S. Automation of Food Processing, Food Engineering. Vol. IV. Wisconsin-Madison: Encyclopedia of Life Support Systems; 2011

[3] Steve. Robotics and Automation for the Food Industry, Trusted Testing for the Food & Feed Industry. Salford: N.P. Analytical Laboratory; 2014

[4] Rachel L. In Praise of Fast Food; 2010. New York: Gastronomica Reader 1st April, 2013

[5] Jijo, Ramesh Kumar. Process automation for a food processing plant. International Journal of Science, Engineering and Technology Research (IJSETR). 2014;**3**(6):1744-1750

[6] Foyose F. The development of a multipurpose wet food sieving machine. Agricultural Engineering International: CIGR E-Journal. 2008;**X**:1-10

[7] Gana IM, Agidi G, Idah PA, Anuonye JC. Development and testing of an automated grain drinks processing machine. Food and Bioproducts Processing, Journal of the European Federation of Chemical Engineering: Part C. Elsevier. 2017;**104**:19-31. www.elsevier.com/locate/fbp

[8] Singh B, Singh BN, Chandra A, Al-Haddad K, Pandey A, Kothari DP. A review of single-phase improved power quality AC-DC converters. IEEE Transactions on Industrial Electronics. 2003;**50**(5):962-981

[9] Abou-elela M, Al-Mashary. Low Cost High Precision Commercial Timer Mode. Riyadh: King Saud University College of Engineering Research Centre; 2005. http://www.college.ksu.edu.su/Engineering/final%20Repot/EE-21-24-26pdf Retrieved on 19th February, 2014

[10] Bimbhra PS, Kaur S. Power Electronics. Vol. 2. Delhi: Khanna Publishers; 2012

[11] Heljo PS, Li M, Lilja KE, Majumdar HS, Lupo D. Printed half-wave and full-wave rectifier circuits based on organic diodes. IEEE Transactions on Electron Devices. 2013;**60**(2):870-874

[12] Rao MB, Daga A, Rao KS, Naveen KM, Kumar BA. Automation of maximum load control using 8051-microcontroller. IOSR Journal of Engineering. 2014;**4**(4):1

[13] Thakur DS, Sharma A. Voice recognition wireless home automation system based on Zigbee. Journal of Electronics and Communication Engineering (IOSR-JECE). 2013;**6**(1):65-75

[14] Ephraim M, Adimula GM, Enelda MA, Ojo OZ, Ibeh SC. Development of a remote controlled vertically adjustable white board. International Journal of Scientific and Research Publication (IJSRP). 2017;**7**(6):717-728

[15] Sharma N, Singha N, Dutta T. Smart bin implementation for smart cities. International Journal of Scientific & Engineering Research. 2015;**6**(9):787-791

[16] Kar R, Sur SN. Object detection for collision avoidance in ITS. European Journal of Advances in Engineering and Technology. 2016;**3**(5):29-35

[17] Agbetuyi AF, Orovwode HE. Design and implementation of an automatic irrigation system based on monitoring soil moisture. Journal of Electrical Engineering. 2016;**6**:206

[18] Gaffa TI, Jideani IA, Nkama I. Traditional production, consumption and storage of Kunu—A non-alcoholic cereal beverage. Journal of Plant Food for Human Nutrition. 2003;**57**(1):73-81

[19] AOAC. Official Methods of Analysis, Association of Official Analytical Chemist. 19th ed. Washington, DC, USA; 2012

[20] Adebayo GB, Otunola GA, Ajao TA. Physicochemical, microbiological and sensory characteristics of kunu prepared from millet, maize, guinea corn and stored at selected temperature. Advance Journal of Food Science and Technology. 2010;**2**(1):41-46

[21] Soya Foods Association of America. Voluntary Standard for the Composition and Labelling of Soya Milk in the United State. 1996. http://www.soyfoods.org/wp-content/uploads/2006/11/smstandards.pdf

[22] Gbabo A, Gana IM, Solomon MD. Effect of blending speed on efficiency and consistency of a grain drinks processing machine. International Journal of Agronomy and Agricultural Research. 2012;**2**(4):16

[23] Nayik GA, Muzaffar K, Gull A. Robotics and food technology: A mini review. Journal of Nutrition & Food Science. 2015;**5**(4):1-4

[24] Jayesh T. 'V-Blender', Lamar Stone Cypher V-76, 841. http/www/brigthub.com/./aspr; 2009. Retrieved 1st April, 2013

[25] Douglas WH. Particle Size Distribution of Manure and By-Product Slurries. US: Oklahoma Cooperative Extension Service; 1997. http://osufacts.okstate.edu

Chapter 9

The Effect of Vermicompost and Other Fertilizers on the Growth and Productivity of Pepper Plants in Guyana

Vasnie Ganeshnauth, Sirpaul Jaikishun,
Abdullah A Ansari and Oudho Homenauth

Additional information is available at the end of the chapter

http://dx.doi.org/10.5772/intechopen.73262

Abstract

Present research was carried out during the year 2014–2015 at the National Agricultural Research and Extension Institute (NAREI) to determine the effect of vermicompost and other fertilizers on the growth and productivity of pepper plants (*Capsicum chinense*). Plants were treated with five different treatments, namely T_1 (Promix), T_2 (vermicompost), T_3 (189), T_4 (189 + vermicompost), and lastly, control which had no fertilizers. T_1, T_3, and T_4 were inorganic fertilizers, and T_2 was organic. Results obtained showed that T_3 (chemical fertilizer) has a significant effect on the growth of pepper plants producing plants with better plant height, number of leaves, number of branches, stem diameter, higher fruit yield, fruit weight and fruit diameter. Plants treated with this treatment also had higher fruit yield, fruit weight, and fruit diameter. Mineral nutrients were highest in plants treated with inorganic fertilizers as compared to the organic fertilizer. Maximum chlorophyll level was present in plants treated with T_2. There were relatively high levels of pest and diseases in plants treated with chemical fertilizers, delayed flowering and fruiting period and high levels of leaf and fruit abscission as compared to plants treated with organic fertilizer (T2). Moreover, T_3 has proven to have a greater effect on the growth parameters of pepper plants but not the quality of plants produce.

Keywords: vermicompost, organic agriculture, chemical fertilizers, plant productivity, pepper

INTECH
open science | open minds

1. Introduction

Peppers (*Capsicum chinense*) belong to the Solanaceae family. They are grown worldwide and are widely appreciated for their spicy flavor and nutritional value. Peppers were usually grown using conventional applications of inorganic fertilizers and pesticides. However, due to the rising awareness of the adverse economic and environmental impact of chemicals in crop productions, the utilization of organic farming has been stimulated as the main farming method today. Organic farming involves the use of organic materials without chemical contributions for growing crops [1].

Organic manures for growing crops are a composition of waste materials. Due to the steady increase in population size and improved living standards around the world, the built up of waste materials is becoming a burgeoning problem since these waste materials emit harmful substances to the atmosphere when burnt. Burning also kills the microbial population of the soil, destroys the soil organic matter, and affects the overall physical composition of the soil [2]. Thus, proper waste management can be maintained by using these organic wastes as substrate in agriculture through organic farming.

Composting of organic waste offers solution to large amounts of waste worldwide. Composting is a natural process of recycling decomposed organic materials into a rich soil known as compost. Traditional composting of organic wastes has been known for years, but new methods of thermophilic composting have become much more common in organic waste treatment [3]. One such composting technique is vermicomposting. Vermicomposting is a type of organic farming by which earthworms breakdown organic waste materials, stimulate microbial activity, and at the same time, increase the rate of mineralization of the soil. These activities convert waste materials into humus-like substances called vermicompost. Vermitechnology is the use of surface and subsurface local varieties of earthworms [4]. Earthworms play a major role in breaking down waste materials to form vermicompost. Vermicomposts are finely divided peat like materials with high water holding capacity, perfect structure, porosity, and aeration. Vermicompost is an organic fertilizer that is rich in nutrients, poor in readily biodegradable carbon, and relatively free of any plant and human pathogens [5]. It has greatly increased surface area, which provides greater area for microbial activity to take place and strong adsorption and retention of nutrients [6, 7].

The activity of organic farming through the use of vermicompost would be an unpreventable practice for years to come for sustainable agriculture, since vermicompost releases nutrients at a slow rate that allows for easy uptake by plants and improves the moisture holding capacity of the soil that results in better quality of crops produce [8]. Ansari [2] outlined different sources of recyclable organic waste, and he classified these waste as either agricultural waste, animal waste, urban solid waste, or agro industrial waste. Animal manure, categorized as animal waste, is a valuable resource as soil fertilizer, since it provides relatively large amounts of macronutrients and micronutrients for crop growth and production and at the same time, providing an environmentally friendly alternative to mineral fertilizers [9].

Heavy use of agrochemicals since 1960s increased food productivity at the cost of environment and society. It killed the beneficial soil organisms, destroyed their natural fertility, and weakened the power of "biological resistance" in crops, making them more susceptible to pests and diseases. Since then, the revolution of vermicomposting studies has been on the go for improving crop production. The use of vermicompost for planting has been highlighted in agriculture as a beneficial medium for improving plant growth and yield and the maintenance of soil fertility. This organic matter has proven to improve the overall soil structure, soil fertility, and crop yield [3]. The aim of this project is to investigate the effect of vermicompost and other fertilizers on the growth and productivity of pepper plants (*C. chinense*).

Organic farming is greatly beneficial and is more economically viable than inorganic farming. Organic farming controls pest and diseases without harming the environment, prevents pollution, and increases soil fertility, so that crops produce will contain adequate nutrients, and better marketable price will be offered. Vermicompost is one of the best organic media for planting. Vermicompost is highly organic and contains no chemicals, so it is environmentally friendly. It is more nutritious and releases nutrients at a slow rate that is easily taken up by plants, and it eliminates the need for application of pesticides, since plants are healthy and free from any pest and diseases. The aim of this research is to determine the effect of vermicompost and other fertilizers on the growth of pepper plants. It will demonstrate how common organic waste can be converted into a nutrient rich substrate that is chemical free and has a massive impact on the quality of crops produce. This research will be of major benefits to farmers in improving their understanding on how vermicomposting can improve the quality of crops produce, increase the fertility of the soil, and reduce the cost needed to purchase synthetic fertilizers for growth, since vermicompost contains all the essential nutrients that support maximum growth. Not only this research will benefit farmers, but also it will benefit the environment by reducing pollution rate, since waste materials can be used as substrate for enhancing soil fertility. Organic farming plays a major role in agriculture today and will be a great influence in the future for safe and good quality of crops. Several researches that were done have proven the importance of vermicompost and its impact on crop production as compared to other fertilizers.

2. Materials and method

Vermicomposting unit was set up at the National Agricultural Research and Extension Institute (NAREI) at Mon repos, Georgetown. All plants were planted at NAREI. Physicochemical analysis and microbial analysis of planting substrates were done at The University of Guyana, Faulty of Natural Sciences Biology Laboratory. The chemical analysis of fruits was done at the Fruit and Drug Department.

2.1. Preparing the vermicomposting unit

1. A vermicomposting unit of dimensions 2.1 × 2.1 × 1 m^3 was set up [4]
2. The floor of the unit was covered with 5 inches of pebbles followed by 10 inches of sand to ensure proper drainage. A 10-inch layer of moisten loamy soil was then placed at the top.
3. 500 locally species of earthworms (*Eisenia foetida)* were introduced into the soil.

4. After inoculation of worms, cattle dungs were scattered over the soil followed by a 10 cm layer of dried grasses and leaf clippings from NAREI Campus. The dried grass along with cattle dung was turned on a weekly basis.
5. After 60 days, the vermicompost was harvested, and the pH was tested and stabilized with calcium carbonate to maintain a neutral pH.
6. The vermicompost was then ready to use as a fertilizer for planting.

2.2. Physicochemical analysis of planting substrates before and after planting

Each planting substrate was subjected to physicochemical analysis, where both the initial soil and soil obtained after planting were analyzed. Planting substrates were analyzed for the following parameters at two different laboratory [10]:

i. pH electrical, conductivity (EC) (done in the Biology Lab at the University of Guyana)

ii. Organic carbon, Nitrogen, Phosphorus, and Potassium (done at Food and Drug Department)

2.3. Microbial analysis

All microbial analysis steps were repeated for each treatment on the initial planting substrate, substrate obtained from seedlings before transplanting to potting media, and substrate obtained after harvesting. Total microbial count was done by culturing microbes on nutrient agar following the procedure as described by Aneja [11]. The modified Winogradsky medium was used for growing and counting *Nitrosomonas* bacterial colonies. Isolation and enumeration of *Azotobacte*r colonies were done using Ashby's medium [11].

2.4. Setting up planting medium

Step 1: Setting up seedling trays (Germination of seeds)

Pepper seeds were planted in a seedling tray of dimensions 53 × 53 cm^2 with a total of 128 holes per tray. The experiment was done following the Randomized Block Design method with three replications for each treatment. Five treatments (**Table 1**) were involved in the replication process.

Step 2: Setting up potting media.

After 4 weeks of growth in seed trays, the seedlings were transplanted into potting media. Each pot was filled with 3 kg of dry soil and 250 g of each treatments were applied to each pots. A total of nine pots were allocated per treatment.

Table 2 shows the amount of vermicompost applied during the different stages of planting. Twelve holes were allocated per treatment, where each set of the 12 holes was filled 50 g of the different planting substrate. Seedlings were planted in each holes and the seed tray was placed in a partially covered area where there was little sunlight penetration and protection from excess rainfall. After 8 days of planting, the seeds have started germinating.

http://dx.doi.org/10.5772/intechopen.73262

Treatments (planting substrate)	Components of each treatment
T_1: Promix (organic)	Canadian sphagnum peat moss, perlite, vermiculite, macro nutrients and micronutrients, limestone, wetting agents, and mycorrhizae.
T_2: Vermicompost (organic)	Loamy soil, cow manure, and dry grasses
T_3: 189 (inorganic)	450 g of sand, 550 g sawdust, 90 g chicken litter, 20 g triple super phosphate (tsp), 8 g urea, 0.013 g of calcium carbonate ($CaCO_3$), and 0.4 g molybdenum potash (MoP)
T_4: 189 + vermicompost (organic and inorganic)	189 + vermicompost components
Control	Black sand

Table 1. Different treatments used in the experiment.

2.5. Growth parameters

The recording of growth parameters began after transplanting seedlings into potting media. Growth parameters such as plant height, number of leaves, and leaf fall were taken on a weekly basis along with observation for any pest attack. After being placed in potting media for 5 weeks, plants were transferred out to the field just before the beginning of flowering. Each plant was planted in bins where field observation was completed. Each planting bins were of dimensions 430 cm length by 90 cm breadth. Four hundred grams of each treatment was applied at the beginning of planting in the field, 150 g at the onset of flowering, and 150 g at the beginning of fruiting. The following analyses were taken in the field trials:

- Number of leaves
- Plant height: measured using a measuring ruler (cm)
- Diameter of main stem: measured using a ruler (cm)
- Number of branch
- Bolting period
- Number of fruits and fruit setting

Treatment	Amount of vermicompost applied (g)				
	Germination	Potting media	Field	Flowering	Fruiting
T1	50	250	400	150	150
T2	50	250	400	150	150
T3	50	250	400	150	150
T4	50	250	400	150	150

Table 2. Amount of vermicompost applied during different stages of planting.

2.6. Application of neem extract to avoid pest

0.6 Kg (600 g) of neem leaves *(Azadirachta indica)* were collected and boiled with 1 liter of water. After boiling, the mixture was diluted with 5 liter of water and mix with 50 ml of soap. The neem extracts were then filled into spray bottles and applied to plants 3 weeks after planting, before transferring to the field, and before flowering and fruiting.

2.7. Harvesting

After harvesting, the following analyses were taken:

- Root and shoot biomass which involve both wet and dry weight
- Shoot length, Number of leaves, diameter of stem, and number of branch
- Total fruit weight, fruit diameter(cm)
- Biochemical analysis of fruit: Fruit samples obtained were dried in an incubator at temperature range 46–50°C and weighed each day, until a constant weight was obtained. After drying, the samples were crushed using a mortar and pestle and stored in a dry place until it was ready for analysis. Samples were analyzed for Vitamins C at the Food and Drug Department following methods outlined by [12]. Samples were also analyzed for Potassium, Sodium, and Phosphorus at the Guysuco Laboratory, LBI.
- Vitamin C and Chlorophyll content.

3. Results and discussion

Plants need nutrients from fertilizers for growth and survival, since most soil does not provide sufficient nutrients for optimum growth. Fertilizers are essential part of modern farming. Fertilizers may be organic or inorganic, and their effect on plant growth depends upon the necessary nutrients they contain. Organic farming is eco-friendly, improves soil fertility, and sustains higher yield. Chemical farming on the other hand has positive effect on crop growth once use in the correct proportion, but intensive use can jeopardize the conservation of soil and invite new problems, which may post health hazard to the environment. Fertilizers in general are essential in modern farming, and the fertility status of the soil is likely to decline unless adequate amount of nutrients is added to the soil.

The aim of this project was to investigate the effect of vermicompost and other fertilizers on the growth of pepper plants. Results obtained are tabularized along with statistical data.

Plants were treated with four different treatments plus a control medium:

T_1: Promix (Inorganic).

T_2: Vermicompost (Organic).

http://dx.doi.org/10.5772/intechopen.73262

T_3: 189 (Inorganic).

T_4: 189+ Vermicompost (Organic+ Inorganic).

Control: Black sand.

Promix is a light-weight, ready-made mixture with high nutrient retention and water holding capacity to support plant growth. It is made up of perlite and vermiculite, which improves moisture and aeration of the soil. Canadian sphagnum peat moss aids in absorption, limestone for pH neutralization, and micro and macro nutrients. Vermicompost, the second treatment (T_2), is a composition of organic matter form from the decomposition of waste product by the action of earthworms. It is an ideal organic manure for better growth and yield of many plants. One hundred and eighty-nine, the third treatment (T_3), is a newly formulated mixture compose of sawdust, sand, urea, TSP, MOP, chicken litter, and calcium carbonate. Sawdust when mixed with these fertilizers provides an ideal medium for plant growth, since these chemicals are weighed and mixed in the correction proportion require for better plant growth and production.

Physicochemical parameters were conducted on both the initial and final planting substrate to determine their physicochemical composition (**Figure 1 (a)–(f)**). For the initial treatment, pH ranges from neutral to alkali for all treatments except for T_1 and control, which was slightly acidic. All pH levels except T3 were within the pH ranges 6.5–7.5, which is the pH that most plant nutrients are optimally available for plant growth, and this pH range is very compatible to plant growth [13]. The electrical conductivity was lowest for control and highest in T_3. Electrical conductivity is a good indication of the nutrient status of the soil. High electrical conductivity means that there are more nutrients present in the soil hence dissolve more ions leading to a high electrical conductivity. Organic carbon was highest in T_1 and lowest in control. Phosphorus and potassium level were highest in T_2 and lowest in control. The control medium was relatively low in all nutrients. Analysis done on postharvest soil was not done on the control substrate, since there was no plant survival in this treatment. The results for postharvest analysis showed that the level of pH increases among all the treatments except for T_3, where there was a decrease in pH level from alkali to neutral. Electrical conductivity decreases among all the treatments with T_2 having the highest conductivity level and T_4 the lowest. There was a decrease in nitrogen, phosphorus, and potassium levels in T_1 and an increase among T_2, T_3, and T_4 with T_3 having the highest level of these macronutrients and T_1 the lowest. Vermicomposts are products from depredated organic matter broken down by earthworms. This process alters the rate of decomposition of organic matter and lowers the C:N ratio [14]. For this reason, vermicompost had low percentage of carbon and nitrogen as compared to the inorganic fertilizer (T_3). Sawdust is a great absorber of nitrogen and absorbs nitrogen from the soil away from plants, and urea is comprised mainly of nitrogen, accounting for the high nitrogen level in T_3. The high phosphorus levels in T_3 are due to the presence of TSP. Moreover, the high levels of macronutrients present in T_3 are due to the chemical composition of the substrate.

Figure 2 shows the results obtained from microbial analysis of both the initial and final planting substrate represented in the form of mean ± standard deviation. The studies of microbial analysis of soil were done before the planting of pepper plants and after harvesting. This

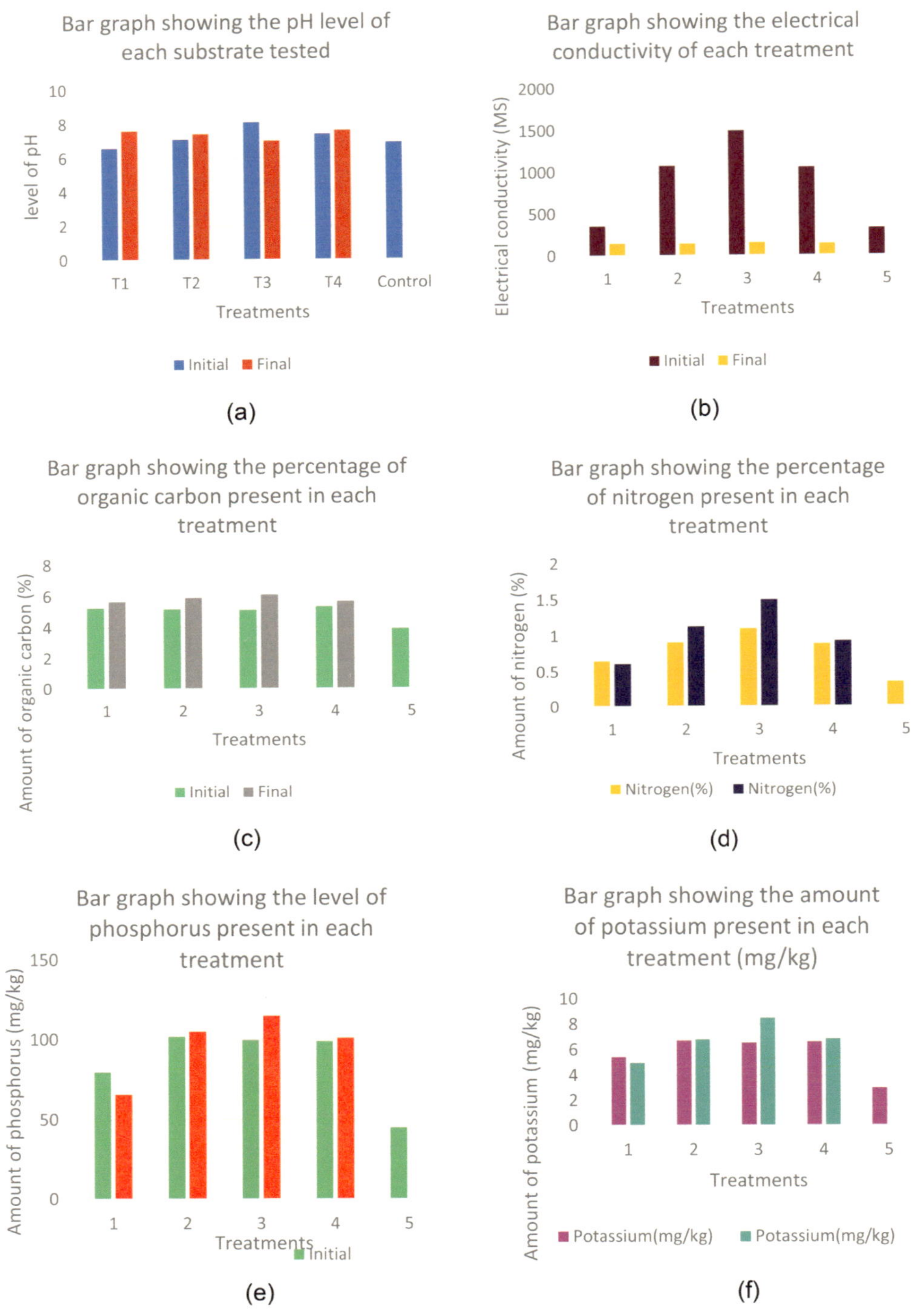

Figure 1. Bar graph showing (a) the pH level of each substrate tested; (b) the electrical conductivity of each treatment; (c) the percentage of organic carbon present in each treatment; (d) the percentage of nitrogen present in each treatment; (e) the level of phosphorus present in each treatment; and (f) the amount of potassium present in each treatment (mg/kg).

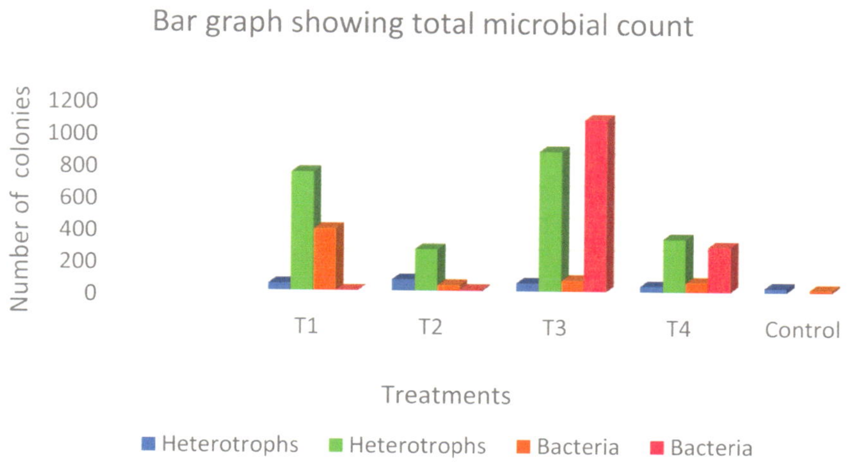

Figure 2. Bar graph showing total microbial count.

gave an idea on the initial microbial count of each substrates and the microbial count after planting. Total microbial count done on the initial soil sample showed that T_2 had the highest amount of heterotrophs as compared to the other treatments, and T_1 had the highest amount of bacteria. For the final microbial analysis, no microbial count was done on the control soil because there was no survival of pepper plants in this treatment. Results obtained from the final microbial analysis showed that T_3 had the highest amount of heterotrophs, as well as the highest amount of bacteria as compared to the other soil samples.

The high microbial count in T_2 for the initial soil sample is due to the presence of microbes deposited from earthworms' casting and microbes naturally present in the soil. Sawdust is rich in fungi, and chicken litter comprises of high amount of bacteria, whereby some might be parasitic but have never shown any effect on human health when amended as a fertilizer for plant growth [15]. This mixture forming the 189 treatment has accounted for the high microbial population present in this treatment due to the continuous application of treatments at different stages of plant growth, which increases the final amount of microbes present in the soil. In addition, the chemical composition of T_3 is acidic, but with the presence of organic matter (chicken litter) and calcium carbonate, the acidity of the mixture is reduced, thereby supporting the growth of more microorganisms [16]. T1 had the second highest heterotrophs for the final microbial analysis, which is due to the presence of mycorrhiza, which is a composition of the promix mixture that creates a symbiotic relationship with plant roots. Statistical analysis done for results obtained on both the initial and final soil sample showed that the results were not statistically significant. Statistical analysis done on the initial soil sample showed that the P-value (0.50) is greater than 0.05 for the treatments and P (0.38) is greater than 0.05 for the different microbes. Analysis done on the final soil sample showed that there was no significant difference between the microbes counted neither between the treatments, since P-value (0.17) is greater than 0.05 for the different treatments and P-value (0.36) is greater than 0.05 for the microbes. After microbial count was done, Gram staining was done on the different bacterial colonies present. All bacteria stained from both initial and final soil sample were Gram-negative rods and cocci.

Figure 3 shows results obtained from nitrifying bacteria through serial dilutions. *Nitrosomonas* and *Azotobacter* are beneficial bacteria that aggressively colonize plant roots and enhance plant growth by a variety of mechanisms which includes phosphate solubilization, antifungal activity, etc. [17]. These nitrogen fixing bacteria are important for the conversion of nitrogen gas to solid nitrogen, which is useable by plants. After total microbial count, serial dilutions were done both on the initial soil samples and final soil samples to determine the total *Azotobacter* and *Nitrosomonas* count in a given amount of soil sample. Dilutions were done at 10^{-3}. Dilutions done on the initial soil samples showed that T_3 had the highest *Azotobacter* count and the second lowest *Nitrosomonas* count. T_3 was followed by T_1 with the second highest Azotobacter count as well as *Nitrosomonas* count,T_2 having the third highest *Azotobacter* count but not *Nitrosomonas* count since T_4 had almost one times more that of T_2, T_4 the fourth highest *Azotobacter* count, and lastly, control with the lowest *Azotobacter* and *Nitrosomonas* count.

For the final dilutions, there was no serial dilution done on the final soil sample for the control treatment, since there were no plants survived. Results obtained showed that T_4 had the highest *Nitrosomonas*, as well as *Azotobacter* count, followed by T_1 with the second highest *Nitrosomonas* count but not the second highest Azotobacter count, since T_3 had one and a half times more of T_1. T_2 had the third highest *Nitrosomonas* count and the lowest *Azotobacter* count followed by T_3 which had the least amount of *Azotobacter*. Overall, for the initial dilution, T_1 had the highest *Nitrosomonas* count and T_3 had the highest *Azotobacter* count. For the final dilution, T4 had both the highest *Nitrosomonas* count and *Azotobacter* count. These results indicated that T4 had good nitrogen fixation process taking place than the other treatments. The results were not statistically significant for the initial soil analysis, since P-value (0.43) is greater than 0.05 between the different treatments, and P-value (0.15) is greater than 0.05 between the different bacteria cultured under serial dilution techniques. Results obtained from the final dilution was not statistically significant neither between the different treatments

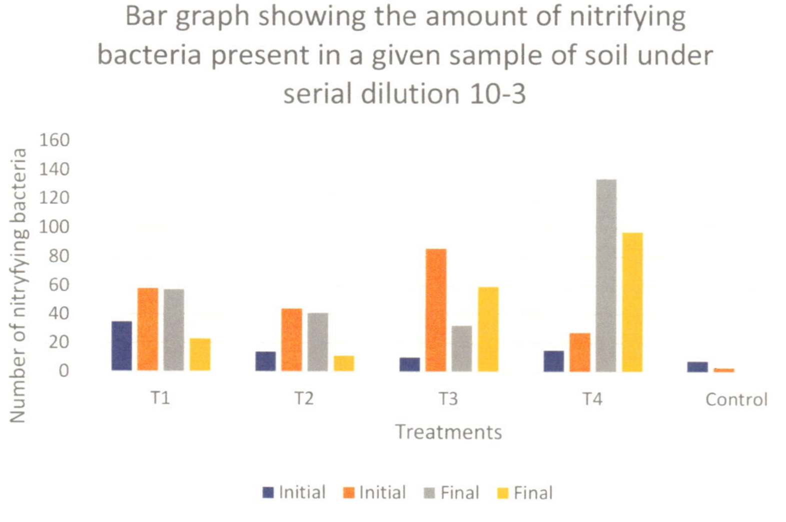

Figure 3. Bar graph showing the amount of nitrifying bacteria present in a given sample of soil under serial dilution 10-3.

nor between the different type of bacteria, since P-value (0.07) is greater than 0.05 between treatments and P (0.29) is greater than 0.05 between the different type of bacteria.

Table 3 shows the rate of germination for different treatments. **Table 4** shows the survival and mortality rate of pepper plants during the different stages of planting. After vermicompost was harvested, all planting substrates were prepared for planting. Pepper seeds were planted on a seedling trays filled with the different treatments, where a total of nine seeds were allocated per treatment. From the germination results, T_2 had the highest germination rate followed by T_1, T_4, T_3, and lastly, control. T_2 has the highest germination rate because vermicompost contains higher amounts of essential nutrients such as phosphorus and potassium which stimulate the emergence of plants [18].

Germination of pepper seeds was followed by transplanting, when seedlings have attained the two leaf stage. Seedlings were transplanted to potting media, where the recording of results began. There was no survival of pepper plants grown in the control medium, so there were no plants to transfer to potting media. This is so because based on results obtained from physicochemical analysis, there was not enough nutrients present in the control medium neither were there enough microbial activity. In the potting media, there was 100% survival of all plants. After transplanting to the field from potting media, there was a change in survival rate among some of the treatments. T_1 and T_4 had 100% survival, 88.89% of plants survived in T_2, and 77.78% in T_3. The change in survival rate is due to the exposure of plants to direct climatic conditions which they were not exposed before. Plants that have died in the field conditions were dry and yellow, which is due to direct contact with the sun. There was a low survival rate in T3 after transplanting to the field. First, this may be due to the pH range of the substrate which was initially 8.1, a pH range where no sufficient nutrients are available for plant growth ([13]). Second, since T3 is composed of sawdust, sawdust as mentioned earlier absorbs nitrogen away from plants, which limits foliage growth causing leaves to yellow and die. T_3 had a much higher electrical conductivity than the other treatments. High electrical conductivity lowers osmotic potential of soil water and consequently the availability of soil water to plants, causing plants to become dry. However, the tolerance of plants to salinity depends upon the plant species, as well as the developmental stages [19].

Figure 4 shows the average plant height obtained from pepper plants grown in five different treatments for a period of 20 weeks. Plant heights were measured on a weekly basis

Treatments	Rate of germination (%)
T_1	83.3
T_2	100
T_3	75
T_4	83.3
Control	41.67

Table 3. Rate of germination in triplicates.

Treatment	Initial amount of plants allocated per treatment	Survival rate in potting media (%)	Survival rate in the field (%)
T_1	9	100	100
T_2	9	100	88.89
T_3	9	100	77.78
T_4	9	100	100
Control	9	0	0

Table 4. Survival and mortality rate of pepper plants planted in each treatment in triplicates.

for 20 weeks. There was a significant difference between plant heights recorded over the 20-week period. T_2 (3.34) had the greatest initial plant height followed by T_1 (2.73), T_4 (2.42), and lastly, T_3 (1.49). The final plant height recorded was higher in T_3 followed by T_4 and lastly T_2 and T_1 with equal average plant height. The percentage change in plant height from initial to final height over the 20 weeks period was greatest in T_3, followed by T_4, T_1, and lastly T_2. The results obtained are similar to results obtained from a study conducted by [20], where the effect of sawdust on the growth of corn was similar to that of pepper. The initial application of sawdust decreases the yield of plant, where in this instance, it decreases the rate of plant growth. This was so because sawdust absorbs nitrogen from the soil away from plants, thus limiting plant growth. As application increases, there was an increase in nitrogen level, which was sufficient enough to cause decomposition and increase the amount of nitrogen available for plant growth. This combined with the amount of nitrogen provided from urea increases the nitrogen level of the soil and subsequently increases the overall plant growth parameters and yield of produce [21]. Results obtained were statistically significant between each treatments, as well as between each weeks of growth, since P-value (0.0059) is less than 0.05 between treatments, and the P-value (3.24×10^{-27}) is less than 0.05 between the different weeks.

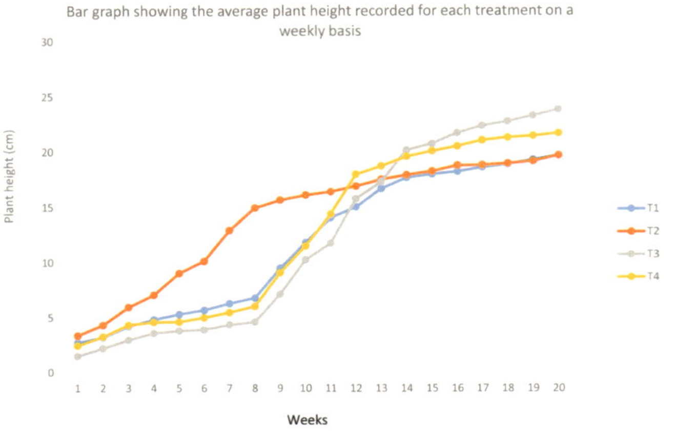

Figure 4. Bar graph showing the average plant height recorded for each treatment on a weekly basis.

Figure 5 shows the average leaf number obtained from pepper plants grown in five different treatments for a period of 20 weeks. Values in the table are represented in the form of mean ± standard deviation (SD). The plants with the greatest overall change in leaf numbers were those grown in T3 followed by T4, T2, and lastly, T1. The treatment with the highest amount of leaves was T_3, which had a sharp increase from week 16 then decreases back at week 19 and 20 but still remained the treatment with the highest average number of leaves. T_3 was followed by T_2, which had a greater overall leaf number than T_1 and T_4, since it started off having a higher leaf number from weeks 1–16, but at weeks 17–20, there was a reduction in leaf number as compared to T_4. T_4 was the next treatment that has plants with a greater leaf number after T_2, where there was a continuous increase in leaf number until the final week. Lastly, T_1 had the lowest number of leaves, where there was a slow increase in leaves until week 20, where it decreases. There was an increase and decrease in leaf number due to leaf abscission. There was a significant difference between each treatment as well as between each weeks, since P-value (0.0016) is less than 0.05 between each treatments and P-value (0.0012) is less than 0.05 between each weeks.

Figure 6 shows the final plant parameters recorded after harvesting of pepper plants from each treatments. The values are represented in the form of mean ± standard deviation. Final plant growth parameters such as plant height, number of leaves, diameter of main stem, and number of branches were greater in plants treated with T_3 followed by T_4, T_2, and T_1, respectively. The results obtained are similar to results obtained from a study carried out by [18] on wheat, where the use of chemical fertilizers has given better growth rate, yield, and quality of produce than vermicompost. Chemical fertilizers have greater availability of salts like nitrate, phosphate, and potash, which significantly increase the rate of plant growth [4]. So T_3 having chemical composition gave better results followed by the mixture of chemical and organic fertilizer (T_4), which had equal proportion of organic and chemical fertilizers to support good plant growth, and then T_2-vermicompost has humic acids and adequate nutrients

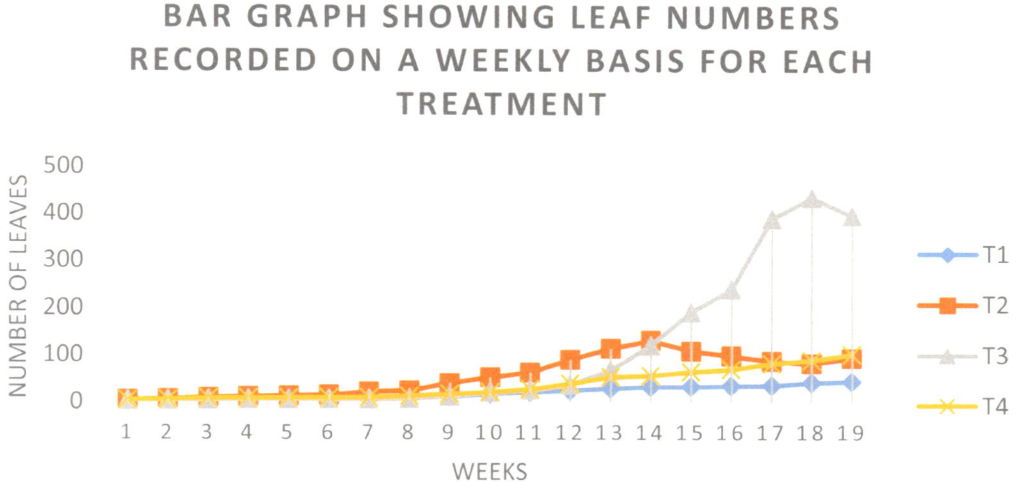

Figure 5. Bar graph showing leaf numbers recorded on a weekly basis for each treatment.

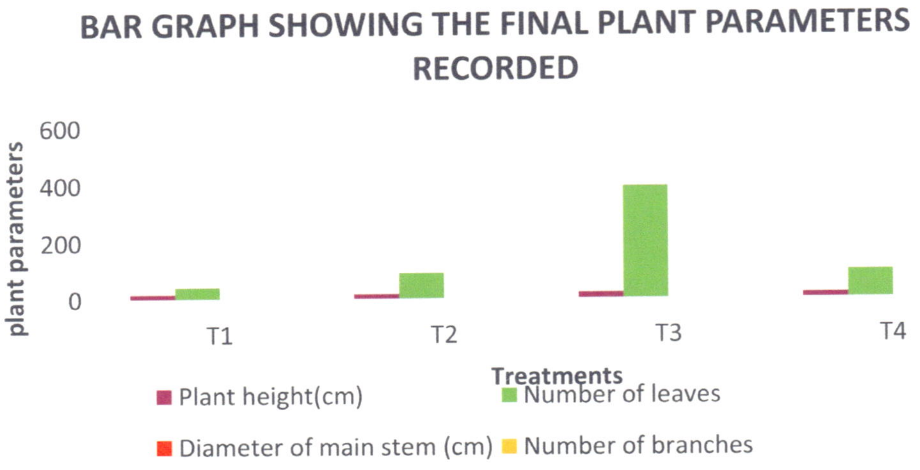

Figure 6. Bar graph showing the final plant parameters recorded.

for maximum growth but not enough micronutrients such as nitrogen, phosphorus, and potassium to produce maximum yield. In addition, since treatments were applied at different stages of planting, [22] concluded from their studies that once vermicompost reaches a certain concentration, the rate of plant growth decreases probably due to the high concentration of soluble salts in the vermicompost, poor porosity, and/or poor aeration. T_1-promix did not have the least average plant growth parameters recorded, which might be due to the presence of insufficient nutrients. Statistical analysis done on the final parameters recorded showed that results were not statistically significant between the different treatments neither between the different parameters recorded, since the p-value was 0.4 between the different treatments and 0.06 between the different parameters recorded.

Figure 7 shows results obtained from analysis of the chlorophyll content of leaves obtained from peppers plants grown in the different treatments. There was a low standard deviation among all the values, which indicates that the values did not deviate much from the mean value. The presence or absence of chlorophyll in plants greatly affects the production of secondary metabolites and other essential plant constituents. In the present study, chlorophyll content in pepper leaves was maximum in T_2 followed by T_3, T_4, and T_1, respectively. Nitrogen is required for cellular synthesis of enzymes, proteins, chlorophyll, DNA, and RNA and is therefore important in plant growth and production of food. Nitrogen fertilization increases growth and leaf area of plants, which in turn increases absorption of light, leading to an increase in the production of chlorophyll [18]. Even though T_2 did not have the highest nitrogen level, it had sufficient to support maximum chlorophyll production followed by T_3, T_4, and lastly T_1, which had the lowest nitrogen level thus the lowest amount of chlorophyll. The results from the ANOVA statistical test showed that there was indeed a significant difference between each treatments as well as the different type of chlorophyll ('a' and 'b') and the total chlorophyll content in leaves obtained from the different treatments. There was a significant difference between each treatment, since F (34.12) is greater than F crit (4.76) and the P-value

http://dx.doi.org/10.5772/intechopen.73262

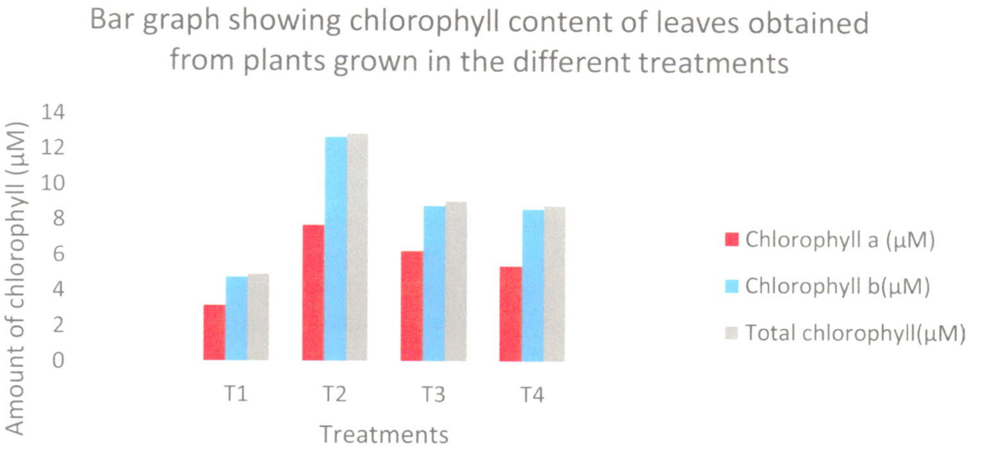

Figure 7. Bar graph showing chlorophyll content of leaves obtained from plants grown in the different treatments.

(0.00036) is less than 0.05, and significant difference between the amount of chlorophyll present, since F (20.08) is greater than F crit (5.14) and the P-value (0.0012) is less than 0.05.

Figure 8 (a)–(b) show the fruit weight and shoot and root weight of pepper plants after harvesting. Results are represented in the form of mean ± standard deviation, where a low standard deviation indicates better results rather than a high standard deviation. The results showed that plants treated with the different treatments had a significant effect on the fresh and dry weight of plants root and shoot. The favorable effect of fertilizer application was most apparent in plants treated with T_3, which had the heaviest fresh and dry shoot weight followed by T_2, T_4 with the third heaviest fresh shoot weight and the lowest dry weight, and lastly, T_1 with lowest fresh weight and higher dry weight than T_4. In terms of the root weight, T_3 also had the highest fresh and dry root weight followed by T_2, T_4, and lastly T_1.

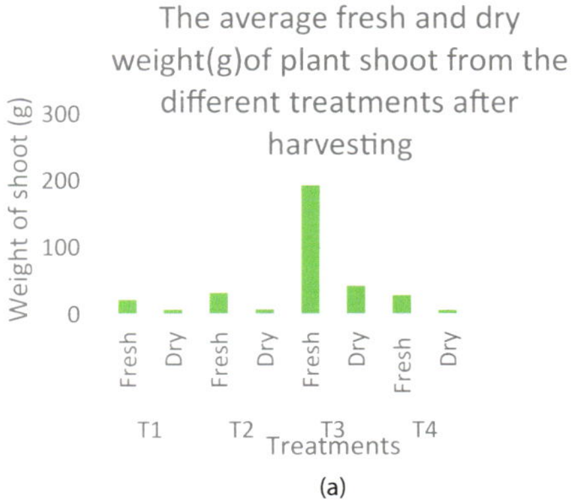

(a)

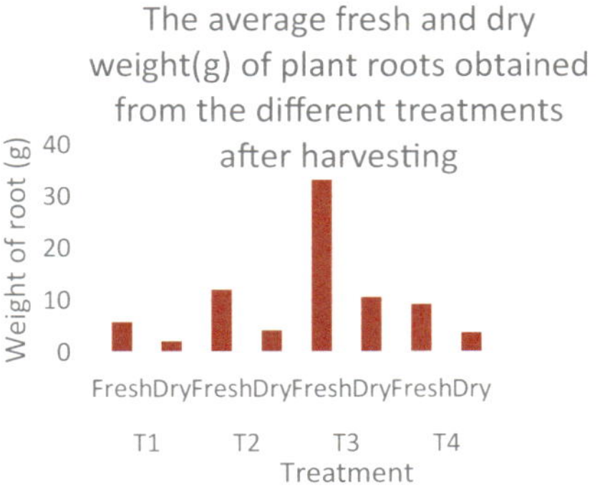

(b)

Figure 8. The average fresh and dry weight (g) of (a) plant shoot from the different treatments after harvesting and (b) plant roots obtained from the different treatments after harvesting.

T_3 had the highest root and shoot weight, which is probably due to the high phosphorus level present in the soil. T_2, vermicompost, had the second highest shoot and fruit weight. According to [23], hormone-like activity of vermicompost leads to an increase in root biomass, root initiation, and better growth and development of plants. Vermicompost is a rich source of humic acids, and humic acid increases plant growth and root biomass [18]. Flowering period for the different treatments varied. T_2 had the earliest bolting period, followed by T_3, T_4, and lastly, T_1. The period of flowering and fruiting also varied among the different treatment, where all treatments except for T_3 have a short fruiting time of 1 week. Addition of vermicompost enhances microbial activity and hence nitrogen levels causing greater root expansion, which in turn leads to greater uptake of nutrients, water, and rate of photosynthesis, ultimately leading to better flowering and heading. For this reason, T_2 had an early flowering period [24].

Table 5 shows the fruiting and flowering period of pepper plants grown in the different treatment. **Figure 9 (a)** and **(b)** shows the fruit parameters of pepper after harvest. Plants amended with T_3 had the highest fruit yield, followed by T_2, T_4, and lastly, T_1. In relation to fruit weight of pepper samples, T_3 had the highest average fruit weight followed by T_2, then T_1 which had only a slightly higher fruit weight than that of T_2, and lastly, T_1 with the lowest average fruit weight of approximately 1 g lighter than that of T_1. Chemical fertilizers have proven to have a better effect on fruit weight and fruit yield in this study.

Figure 10 shows results obtained from nutrient analysis of pepper samples grown in the different treatment. **Figure 11** shows the vitamin C content in fruit samples obtained from different treatments. Pepper samples were analyzed for their nutrient content.T_2 and T_4 had maximum amount of sodium, T_3 had the highest percentage of potassium, and T_3 and T_1 had the highest percentage of equal amounts of phosphorus. Vitamin C amount was highest in T4, which is a combination of organic and inorganic fertilizer. This combination has proven good results on the nutritional value of pepper. This was followed by T_3, T_2, and lastly, T_1. Collectively, all results obtained have favored T_3 (chemical fertilizer). Plants response to T_3 (chemical fertilizer) is better than any other planting medium for growing pepper plants, since it is rich in nutrients and microbes, which increase plant height, leaf numbers, and number of branches and stem diameter. It also increases the fruit yield and nutrient quality of fruits produce. However, despite these positive effects on pepper plants, there were presence of pest (whiteflies) and diseases on all pepper plants grown in T_3. Neem extracts were sprayed on all pepper plants, since neem is known as a natural insect repellent. However, plants

Treatment	Flowering period	Fruiting period
T_1	Week 16	Week 17
T_2	Week 8	Week 9
T_3	Week 12	Week 14
T_4	Week 13	Week 14

Table 5. Flowering and fruiting period of pepper plants.

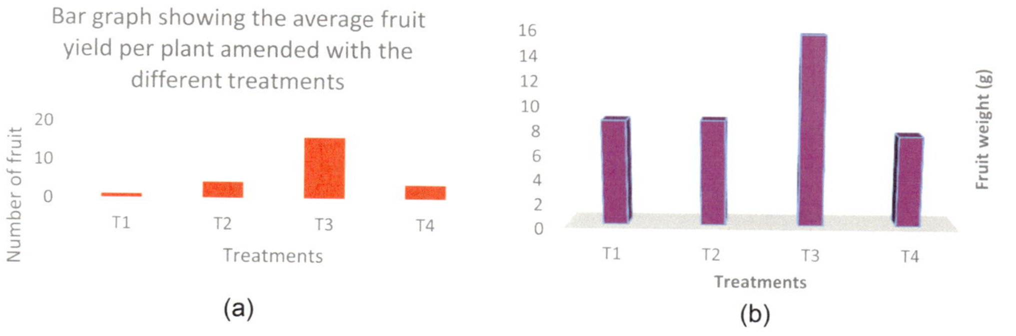

Figure 9. Bar graph showing the average fruit yield per plant amended with the different treatments.

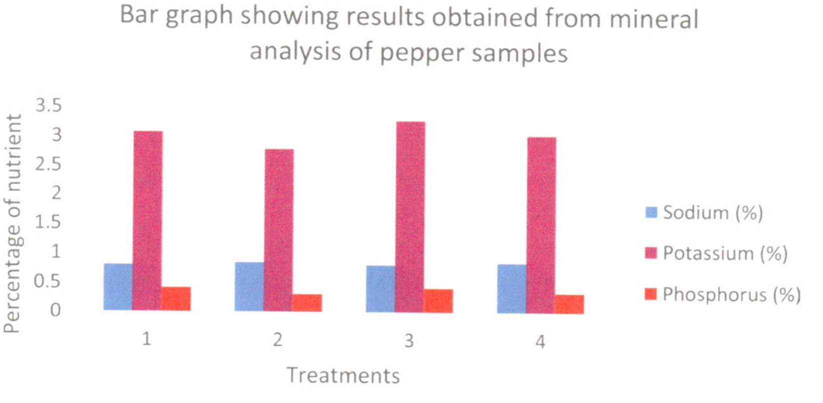

Figure 10. Bar graph showing the results obtained from mineral analysis of pepper samples.

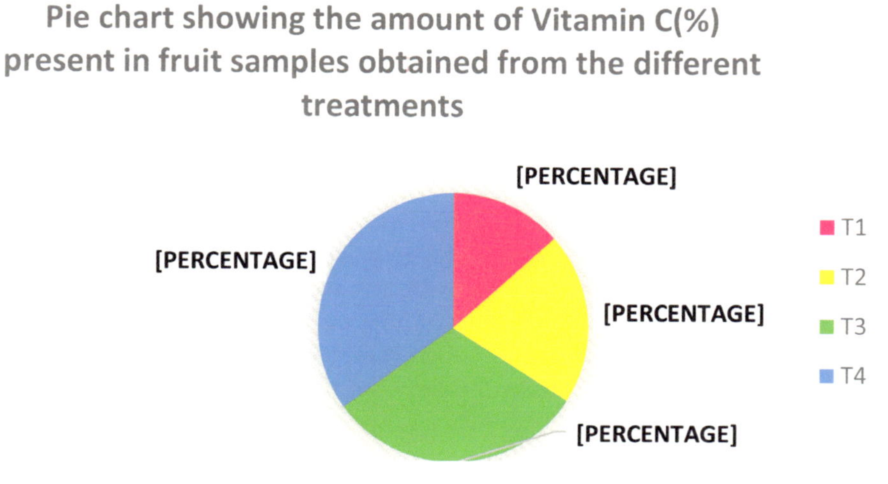

Figure 11. Pie chart showing the amount of vitamin C (%) present in fruit samples obtained from the different treatments.

grown in chemical fertilizers were still susceptible to pest and diseases. Ref. [16] stated that chemical fertilizers increases plant diseases, because they have a higher nitrogen content than slow-release organic fertilizers. With high abundance of nitrogen and phosphorus, plants are susceptible to mosaic infections. Lack of trace elements is also related to fungal and bacterial diseases in plants and vegetables. In addition, even though fruit yields and leaf numbers were high in T_3, there was massive leaf and fruit abscission occurring, which may be due to the hormonal imbalance in plants grown on this treatment [25]. Mineral nutrient applications could cause stimulation of vegetative growth during the period critical to fruit retention resulting in increased fruit drop and loss of yield [26]. T_2 (vermicompost) was the second best treatment for growing pepper plants producing plants with significantly high amounts of chlorophyll as compared to the other treatments, good nutrient content, and faster plant growth rate. Plants treated with T_2 had high growth rate when they were in potting media. However, after transplanting to the field, the rate of plant growth after a period decreases. The reason for this may be due to excessive application of vermicompost, since too much vermicompost limits plant growth [22]. In contrast to T_3, there was no presence of pest and disease attack in this treatment. This is similar to result obtained from a study conducted by [27], where plants treated with vermicompost did not show any signs of pest and diseases, which may be due to the pesticide action of vermicompost that aids in protecting crop plants against pest and diseases by suppressing, repelling, or by inducing biological resistance in plants to fight them. The next treatment, T_4, a mixture of T_2 and T_3 (organic and inorganic), was proven as the third best treatment for growing pepper plants with moderate plant growth rate, good fruit yield, and good nutritional value. However, there was presence of whiteflies and diseases similar to that of T_3. Lastly, T_1 (promix) had little effect on the growth and productivity of pepper plants, even though it had moderate amount of nutrients and there was small amount of diseases present. One reason for the limitation of plant growth in T_1 may be due to the pH level which was acidic, having a negative effect on the microflora population in soil, decreasing nutrient recycling and soil aeration.

4. Conclusion

The use of vermicompost for growing pepper plants did not have a greater effect on plant growth and productivity than other fertilizers. Chemical fertilizers (T_3) have proven to be the best medium for growing pepper plants producing plants with greater plant height, leaf number, number of branches, and fruit yield. Chemical fertilizers not only does affect plant growth positively but also have negative impacts on pepper plants by causing pest and diseases on every plants grown in this treatment and premature dropping of fruits. Pepper plants also had a delay in flowering and fruiting period as compared to vermicompost, and survival rate was negatively affected when compared to the other treatments. With presence of pest and diseases, plants will require pesticide which in turn might leave residue in plants fruits and eventually cycle into our system upon consumption. T_2 was the second best medium for growing pepper plants producing plants with maximum chlorophyll content, faster germination rate and faster growth rate. Treatment T_4 was second best medium producing pepper with high amount of vitamin C whereas in control growth rate of pepper plants was relatively very poor.

Author details

Vasnie Ganeshnauth[1], Sirpaul Jaikishun[1], Abdullah A Ansari[1]* and Oudho Homenauth[2]

*Address all correspondence to: abdullahadilansari@gmail.com

1 Department of Biology, Faculty of Natural Sciences, University of Guyana, Turkeyen Campus, Georgetown, Guyana

2 NAREI, Guyana

References

[1] Arancon N, Edwards C, Bierman P, Metzger J, Lee S, Welch C. Effects of vermicompost on growth and marketable fruits of field grown tomatoes, peppers and strawberries. Pedobiologia. 2003;**47**:731-735

[2] Ansari A. Vermitechnology- Permutations and Combinations of Organic Waste. LAP LAMBERT Academic Publishing Gmbh and Co. KG; 2012. pp. 6-8

[3] Arancon N. and Edwards C. 2005 .Effects of Vermicompost on plant growth, International Symposium Workshop on Vermi-Technologies for Developing Countries (ISWVT 2005), Los Banos, Philippines. pp. 2

[4] Ansari AA, Sukhraj. Effect of vermiwash and vermicompost on soil parameters and productivity of okra (*Abelmoschus esculentus)* in Guyana. African Journal of Agricultural Research. 2010;**5**(14):1794-1795

[5] Dominguez J, Edwards C. Chapter 2: Relationship between Composting and Vermicomposting. Taylor and Francis Group LLC. 2011. pp. 20-21

[6] Ansari AA, Ismail SA. Management of Organic Waste: Earthworms and Vermiculture biotechnology. Management of Organic Waste (Chapter 5). 2012a:90-92

[7] Ansari AA, Ismail SA. Role of earthworms in Vermitechnology. Journal of Agricultural Technology. 2012b;**8**(2):403-415

[8] Najar I, Khan A. Effect of vermicompost on the growth and productivity of tomato (*Lycopersicon esculentum*) under field conditions. Acta Biologica Malaysiana. 2013;**2**(1): 12-21

[9] Lazcano C, Dominguez J. The Use of Vermicompost in Sustainable Agriculture: Impact of Plant Growth and Soil Fertility (Chapter 10). ISBN: 978-1-61324-785-32011. p. 1

[10] Homer F. Soil Analysis Manuel. Guyana: Central Analytical and Environmental Monitoring Services, Agriculture Research Department, LBI; 2003. p. 100

[11] Aneja KR. Experiments in Microbiology, Plant Pathology, Tissue Culture and Mushroom Cultivation. Second ed. New Delhi: Wishwa Prakashan (New Age International (P) limited); 1996 pp: 36-38, 64-68, 118-120

[12] Thangjam R. Biochemical Methods. Second ed. New Delhi: New Age International (P) Limited; 1996. pp. 178-186

[13] International Plant and Nutrition Institute (IPNI), 2015. Soil pH and the Availability of Plant Nutrients [Internet]. The Fertilizer Institute [Cited 2015-07-27]. Available from: <http://www.nutrientstewardship.com/implement-4rs/article/soil-ph-and-availability-plant-nutrients>

[14] Mathiavanan S, Kalaikandhan R, Chidambaram ALA, Sundramoorthy P. Effect of Vermicompost on the growth and nutrient status in groundnut *(Arachishypogaea. L)*. Asian Journal of Plant Sciences and Research. 2013;**3**(2):15-22

[15] Wilkinson KG, Tee E, Tomkins RB, Hepworth G, Premier R. Effect of heating and aging of poultry litter on the persistence of enteric bacteria. Poultry Science. 2011;**90**:10

[16] Monroe. Organic Fertilizer Vs. Chemical Fertilizer-Does it Matter? [Internet]. 2015 Monroe Works [Cited July 20th, 2015). Available from: <http://www.monroeworks.com/organic-fertilizers-vs-chemical-fertilizers.html>

[17] Ibiene AA, Okerentugba PO, Akhigbemen OJ. Effect of some plant growth promoting rhizobacteria (PGPR) on growth of Piper Nigra. Stem cell. 2014;**4**(4):17-26

[18] Joshi R, Singh J, Vig AP. Vermicompost as an effective organic fertilizer and biocontrol agent: Effect on growth, yield and quality of plants. Reviews in Environmental Science and Bio/Technology, ISSN: 1569-1705. 2014:1-25

[19] Patel AD, Lalcheta K, Gill SS, Tuteja N. In: Tuteja N, Gill SS, editors. Salinity Tolerance of *Avicennia officinalis L.* (Acanthaceae) from Gujarat Coasts of India. Gemany: Wiley; 2014. pp. 1-20

[20] Locascio SJ, Nettles VF, Neller JR. The Effect of Sawdust Incorporation in a Soil on Growth of Irish Potatoes and Sweet Corn. Gainsville: Florida Agricultural Experiment Station; 2008. pp. 197-201

[21] Iderawumi AM, Olusola OS, Friday CE. Integrated application of urea and sawdust ash: Effect on soil chemical properties, plant nutrients and sorghum performance. IOSR Journal of Agriculture and Veterinary Science. 2012;**1**(4):38-41

[22] Chamani E, Joyce DC, Reihanytabar A. Vermicompost effects on the growth and flowering of Petunia Hybrida dream neon rose, American-Eurasian. Journal of Agricultural and Environmental Science. 2008;**3**(3):506-512

[23] Bachman GR, Metzger JD. Growth of bedding plants in commercial potting substrate amended with vermicompost. Bioresource Technology. 2007;**99**:3155-3161

[24] Taleshi K, Shokoh-far A, Rafiee M, Noormahamadi G, Sakinejhad T. Effect of vermicompost and nitrogen levels on yield and yield component of safflower (Carthamus Tinctorius L.) under late season drought stress. International Journal of Agronomy and Plant Production. 2011;**2**(1):15-22

[25] Ashraf MY, Ashraf M, Akhtar M, Mahmood K, Saleem M. Improvement in yield, quality and reduction in fruit drop in Kinnow (citrus Reticulata Blanco) by exogenous application of plant growth regulators, potassium and zinc. Pakistan Journal of Anatomy. 2013;**45**:434-439

[26] Lovatt CJ. Factors Affecting Fruit Set/Early Fruit Drop in Avocado [Internet]. California Avocado. Society 1990Yearbook [Cited 2015, July 14]. Available from: <http://www.avocadosource.com/CAS_Yearbooks/CAS_74_1990/CAS_1990_PG_193-199.pdf>

[27] Mamta, Wani K, Rao R. Effect of vermicompost on growth of brinjal (*Solanum melongena*) under filed conditions. Journal on New Biological Reports. 2012;**1**(1), ISSN: 2319-1104: 25-28